FOOD SAFETY AND THE AGRO-ENVIRONMENT IN CHINA: THE PERCEPTIONS AND BEHAVIOURS OF FARMERS AND CONSUMERS

Edited by **Teruaki Nanseki**
and **Min Song**

Food Safety and the Agro-Environment in China - The Perceptions and Behaviours of Farmers and Consumers
http://dx.doi.org/10.5772/56022
Edited by Teruaki Nanseki and Min Song

Contributors
Dr. Teruaki Nanseki*, Professor, Faculty of Agriculture, Kyushu University, Japan
Dr. Min Song, Professor, Institute of Agricultural Resources and Regional Planning, Chinese Academy of Agricultural Sciences, P. R. China
Dr. Tinggui Chen, Associate Professor, College of Economics and Management, Shanghai Ocean University, P. R. China
Dr. Dongpo Li, Assistant Research Fellow, Institute of Rural Economics, Hebei Academy of Social Sciences, P. R. China
XiaoouGao, Assistant Research Fellow, Institute of Agricultural Resources and Regional Planning, Chinese Academy of Agricultural Sciences, P. R. China
Dr. Hui Zhou, Assistant Research Fellow, Institute of Agricultural Economic and Development, Chinese Academy of Agricultural Sciences, P. R. China
* Corresponding author (e-mail: nanseki@agr.kyushu-u.ac.jp)

Published by InTech
JanezaTrdine 9, 51000 Rijeka, Croatia

Publishing Process Manager Danijela Duric
Typesetting www.pantype.com
Cover InTech Design Team

Additional hard copies can be obtained from orders@intechopen.com

Food Safety and the Agro-Environment in China - The Perceptions and Behaviours of Farmers and Consumers, Edited by Teruaki Nanseki and Min Song
p. cm.
ISBN 978-953-51-1000-2

Meet the editors

Dr. Teruaki Nanseki is a professor of Kyushu University, Japan. His research interests cover management, informatics and operational research in agriculture. Management of risk, information and talent is the major research topic in his ongoing national research projects. He received B.S. in Agronomy and M.S. in Farm Management from Okayama University in 1980 and 1982 respectively. He was a visiting fellow at Cornell University from 1980 to 1981 and received Ph.D. in Agricultural Economics from Kyoto University in 1990. He has published more than 300 publications in scholarly journals, books, conference proceedings and academic reports. He has received seven academic prizes from the academic societies.

Dr. Min Song is a professor of Institute of Agricultural Resources and Regional Planning, Chinese Academy of Agricultural Sciences, P. R. China. He mainly engages in researches on resources and environment economics, intellectual property rights and public policy in agriculture. He received B.S. and M.S. in Agricultural Economics from Southwest Agricultural University of China in 1985 and 1993 respectively. He was conferred Ph.D. in Agricultural & Resource Economics from Kyushu University in 2002. He has published more than 100 publications in scholarly journals and books, and received many prizes for the supervision of research projects, and offering policy suggestions etc.

Contents

Preface

Is the 21st century going to be Asia's century, especially of China? The world is greatly concerned by this question, which is difficult to answer. According to Sommer (2006) [1], *"A power shift from the West to the East is in the offing. A new world order is taking shape. The 21st century is still young. The Americans are hoping for a second American century. But perhaps – given the rise of Asia, especially China – they will have to settle for an American half century. Only one thing is certain: the second awakening of China will bring about a power shift in the world."* This prediction of China's rise is the subject of this book.

When we ponder the answer to the question, we need to consider important facts stated in Sommer (2006) [1]: *"Asia was the birthplace of the three Abrahamic religions. Epoch-making inventions and discoveries had reached the West from there... but this was only one side of the coin. Simultaneously, Asia was seen as static, inert, immobile... the Asia-Pacific region has become a model of modern development dynamics. Japan started first."*

Is the environment an important issue in Asia? Kyushu University established the Research Institute for East Asia Environments (RIEAE) in 2009. The president of the universitystated the reasons for establishing the institute in Arikawa (2009) [2]: *"Currently, the entire world, especially the East Asian region, is facing serious environmental damage. This damage involves various issues including air pollution, river and marine pollution, waste problems, and food contamination—issues too numerous to innumerate in full—and which is far beyond an individual researcher's capability to address. Integrating relevant members of the university as an organization through the cooperation of relevant research institutes as well as the support of private companies, RIEAE is conducting a range of activities with an eye on making a social contribution at the national level to practically solve these environmental issues, which have become complicated."*

What is the contribution of this book? The main idea of this book is *Kanshokufuji*—環食不二 in Japanese and 环食不二 in Chinese. This new concept proposed in Nanseki (2008) [3] pertains to a sound food system in a sound agro-environment. It implies that food supply and demand are non-separable in terms of location due to the importance of retaining the suitable condition of the environment and biodiversity in the region. This concept is necessary and useful for coping with environmental issues and related food safety issues.

Food science, considered as a branch of agricultural science in its broad meaning, has made valuable contributions to food safety. However, in the real agro-food chain, agricultural products may be contaminated by environmental pollution, to which agricultural production may contribute. This dilemma implies that contemporary science is still missing the point of an agro-environment for crops and livestock in that agriculture production including agrochemical utilization, food processing and marketing, food consumption, and residue processing need to be viewed as a whole system. In this sense, the concept of *kanshokufuji* is also useful and needed in environmental science to prevent environmental damage as well as enhance food safety.

From this view point, several integrated surveys in both rural and urban areas of China were conducted by the food risk research group at the RIEAE to reveal the current status of the environment, food, and agriculture in this country. The results of the surveys are introduced in this book, along with implications and recommendations. This book presents the beginning of our research, and we are happy to receive comments from our readers. Finally, the authors gratefully acknowledge the financial support of the RIEAE, Kyushu University.

Teruaki Nanseki
Faculty of Agriculture,
Kyushu University,
Japan

Min Song
Institute of Agricultural Resources and Regional Planning,
Chinese Academy of Agricultural Sciences,
P. R. China

References

[1] Sommer T. Is the 21st century going to be the Asian century? ASIEN 2006; No.100: 70-78 (http://www.asienkunde.de/).

[2] Arikawa S. Message from the President, Kyushu University, Research Institute for East Asia Environments 2009: http://www.q-eaep.kyushu-u.ac.jp/message/en/.

[3] Nanseki T. Perspective of information technology for food safety. Japanese Journal of Agricultural Information Research 2008; 17 (4): 161-170. (in Japanese)

Chapter 1

Introduction

Min Song and Teruaki Nanseki

1.1 Problem statement

The Industrial Revolution which began in Britain in the mid-1700s, and spread to the rest of the world around the mid-1800s, not only brought the rapid development of the industrial economy, but also led to the expansion of numerous environmental hazards. Nevertheless, the negative effects of the Industrial Revolution on the environment were not revealed until 1962 in the globally acclaimed book, *Silent Spring*, written by Rachel Carson. In this book, she took on the environmental and human dangers caused by indiscriminate use of pesticides. "Over increasingly large areas of the United States spring now comes unheralded by the return of birds, and the early mornings are strangely silent where once they were filled with the beauty of bird song" (from *Silent Spring*). The agro-environment is an important part of the natural environment and the basic material condition for agricultural production. Agro-environment degradation includes ecological destruction and environmental pollution, and the latter is the theme researched in this book. In addition to constraining the sustainable development of agriculture, agro-environmental deterioration also increases risks in food through material recycling. Nowadays, food safety and the agro-environment have become challenges around the world.

China, as a large agricultural country, has a 7,000-year history of sericulture and a 6,400-year history of rice farming. Over a long time, Chinese farmers have engaged in environment-friendly agricultural production, which has had significant impact on present organic agriculture. However, due to food shortages, China introduced industrial agriculture in the 1960s to improve crop yields. Over the last few decades, with a rising industry and economy, and promoted by incentive policies, chemical products have been increasingly put into agriculture in China (Figure 1.1).

Nowadays, China is the largest user of fertilizer, pesticides and plastic film in the world [2]. In 2010, Chinese agriculture consumed 55.61 million tons of chemical fertilizer, 1.75 million tons of pesticide and 2.17 million tons of plastic film [3-4], which is much higher than the world average. As nearly 60%-70% of chemical fertilizers, 60% of pesticides and 40% of plastic film

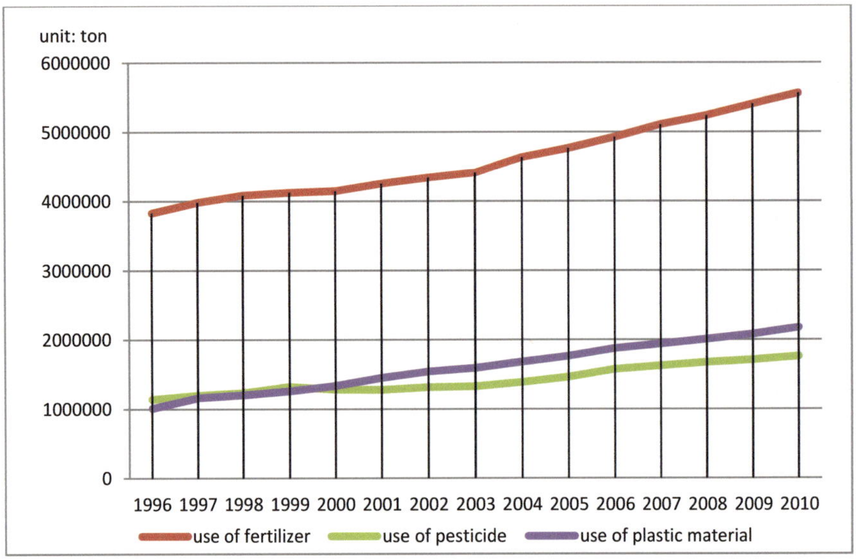

Source: China Ministry of Agriculture (2012) [1]

Figure 1.1 Use of fertilizer, pesticides and plastic material in China

gets in the environment [5-6], excessive use of chemical products is bound to have a notable influence on the ecological environment. In addition, with the development of animal husbandry, animal manure has become another source of agro-environment pollution. According to the data of the 1st national census on pollution sources in China (2010), agricultural production has seriously polluted the air, soil and especially the water environment. Agricultural sources provided 40% of chemical oxygen demand (COD), making a higher contribution than industrial and domestic sources. Simultaneously, agriculture is the major source of emissions of total nitrogen and total phosphorus, accounting for 57.2% and 67.4% of the total emissions, respectively. Agro-environment pollution has become an urgent challenge in China.

As is known to all, an unhealthy agro-environment cannot supply safe crops. With the degradation of the agro-environment in China, a series of food safety incidents broke out in recent years, such as the Hainan drug cowpea incident of 2011, the Qingdao drug leak incident of 2010 and the Guangdong drug watermelon incident of 2007. But agro-environmental pollution is not the only reason, abuse of addictives, microbe contamination and illegal processing operations also gave rise to successive food safety crises in China, e.g., excessive melamine in milk powder, salted duck egg containing Sudan Red and clenbuterol poisoning.

The food safety crises, which have been listed in the top 10 of issues concerning peoples' livelihoods in China since 2005, have brought about a series of severe consequences as follows: first of all, significant outbreaks of food borne disease fundamentally undermine public trust. According to the Report on Chinese Food Safety in 2011-2012 released by the Chinese magazine *Well-off* and Tsinghua University, it was reported that 63.7% of the respondents believed that food safety in China is bad and 80.4% of the respondents thought food is not safe at all in China. Secondly, successive food safety incidents have a negative impact on food exports, weakening the international competitiveness and market reputation of food made in China. The massive media exposure of these food safety incidents has greatly reduced foreign consumers' confidence in food made in China. A survey carried out in Korea shows that nearly 90% of respondents think food imported from China is not safe [7], resulting in the international market raising barriers to limit food imports from China. For example, a number of countries took strong inspection measures on related products made in China after media exposure of a poison capsule event. Thirdly, food safety accidents often result in significant economic losses. Take the melamine incident as an example, it led to the bankruptcy of Sanlu milk enterprise, and at the same time, the entire milk industry suffered a big shock. During the incident, vast amounts of milk were poured away and a great number of dairy cows were slaughtered.

The object of the book is to reveal the challenges of the agro-environment and food safety in China from different perspectives, and try to find some solutions by analysing vast amounts of data gathered by large-scale surveys.

1.2 Literature review

The agro-environment plays a decisive role in developing sustainable agriculture and providing safe food. With increasing industrialization, urbanization and agricultural modernization in China, excessive chemical materials are being widely applied to agricultural production [8]. Thus, the agro-environment is getting worse, which has hampered the development of sustainable agricultural. Economist first focused on the environment in 1920 when Pigou analysed pollution problems in his book on welfare economics. He pointed out that externality was the root of pollution and sewage enterprises should be taxed. For a long time "externality theory" was the overwhelmingly mainstream theory in environmental science. Later, with the development of institutional economics and information economics, economic theories about the environment were greatly promoted. Domestic scholars studied the agro-environment mainly based on the natural mechanisms, and from the technical and engineering perspective. Only a few researchers analysed the agro-environment from the view of economics, which could be divided into following: (1) the microeconomics perspectives. some studies revealed that externality resulted in agro-environmental problems. On the one hand, it is difficult to define the property rights of agro-environment, which will cause "market failure". On the other hand, a lack of incentive and restraint mechanisms with regard to agro-environmental protection will lead to "government failure" [9]. Other scholars proved that in addition to the lack of protection systems, incentive and restraint mechanisms were the root of agro-environment problems through game analysis [10-11]. (2) Institutional economics perspective. Institutional economists considered that the dual control system had resulted in the generation and degradation of rural non-point source pollution [12]. (3) Farmers' behaviour analysis. Zhu (2000) [13] carried a survey in rural Beijing and the results showed that farmers lacked knowledge, awareness and motivation to environmentally protect. Zhu et al (2009) [14] studied farmers' perceptions on the environment through a survey in Hunan province and authors found that awareness among farmers in various regions is correlative with the region's development, and farmers' awareness is not in accordance with their behaviour. According to the results of a survey in Hunan province, Yan (2011) [15] pointed to the fact that farmers' awareness of protecting the environment was increasing.

Researching on food safety using economic methods began in the 1960s. During this period, economists brought forward a series of models to study realities in economic theory. Fortunately, food safety is one of the subjects investigated [16]. In the late 1980s, due to the occurrence of mad-cow disease (BSE), increasing numbers of people began to pay attention to food safety issues. In the 1990s, following the publication of *Economics of Food Safety* at the proceedings of the American National Workshop on the same title held in Washington, a series of reports were published, thus laying the foundation for economic analysis of food safety. Now, more and more economists are studying food safety issues using empirical analysis methods, especially in the study of consumer behaviours, awareness and willingness to pay for safe food. For example, Chern et al (2002) [17] studied consumers' willingness to pay for genetically modified vegetable oils through carrying out a survey in Shikoku, Japan, Norway, Taiwan and the United States. Georges et al (2006) [18] selected objects from 12 European countries and divided them into four groups to understand their perception on the food traceability system by group discussion.

The vast majority of Chinese economic research on domestic food safety issues uses foreign theories and research methods. Wang (2003) [19] carried out a survey of 289 consumers in Tianjin and analysed the process and characteristics of selecting safe food. Chern et al (2002) [17] revealed that consumers surveyed in Zhejiang province were relatively concerned about vegetable safety and their attitudes were found to be negative. Consumers were very willing to pay the extra cost for safe vegetables, but the price of the safe vegetables should be higher than ordinary vegetables by no more than 10% to 20%. Zeng et al (2008) [20] studied consumers' willingness to pay for moon cake with safe additives through a survey of 396 consumers in 25 supermarkets in Beijing.

Research on the agro-environment and food safety respectively from the perspective of farmers and consumers has increased in recent years. Most researchers only analyse and expound survey data, however, in-depth and comprehensive analysis, and related solutions are scant.

1.3 Theoretical framework and methodology

1.3.1 *Kanshokufuji*: The theoretical basis

The authors of this book were driven to develop the contexts based on the important concept of *Kanshokufuji*, a Japanese term created by Prof. Nanseki Teruaki, one of the chief editors of this book. With the literal meaning of *integrated environment and food*, this concept advocates a sound food system in a sound agro-environment [23]. In modern society, food safety is increasingly becoming of global concern, due to asymmetric information on the processes, additives in the long industrial chain, etc [21-22]. In this key concept for establishing a safe and sustainable next generation food system, food is defined using a broader concept and recognized as being supplied via more systematic processes. In addition to processed products ready for eat and drink, food includes the raw materials with plant or animal origins, i.e., any substance consumed to provide nutritional support for the human. Meanwhile, the processes start from the growing condition of the agricultural, livestock and marine production, and cover the following processing and circulations of food. The processes also include eating and drinking, intake and metabolism of food, management of the residues and wastes. Based on the expanded and consistent concepts of food and supply processes, the safety and risks of food need to be studied from various aspects, including food, agriculture and environment [23].

As shown in Figure 1.2, the concept of *Kanshokufuji* aims to establish a safe and sustainable food system, through demonstrating the key notion that a sound food system can only begin from a sound agro-system. Once the soil and water are contaminated by human activities, including agriculture, it is difficult to conduct safe agricultural and livestock production. In the process of agricultural and livestock production, pesticides and veterinary drugs should by applied properly. Nevertheless, safe production of food cannot be maintained if the living environment of the agricultural and livestock products is polluted by cadmium and other heavy metals or poisonous chemicals contained in industrial liquid waste, despite endeavours in the proper application of the means of production. Thus, with this awareness in mind, the assurance of food safety and environmental protection is interpreted as integrated in, or inseparable from, *Kanshokufuji*. In other words, it calls on the technical advances and institutional designs to prevent pollution of the living environment and food stuffs. Meanwhile, there is another important concept of *Ishokudogen* which targets the risks which occur after food enters the body. Based on the latest advances in studies on food functions and inner metabolism, this Japanese term argues that medicine and food are both important to maintain human health, and they are from the same source.

Hence together with *Ishokudogen*, *Kanshokufuji* provides a key notion of establishing a food system covering different generations. Considering the risks of food before entering the body, safety of soil and water constitutes the prerequisite of the overall food safety system. Meanwhile, proper eating and drinking habits are essential in maintaining health. Thereafter, when designing the cross-generation food system, it is of great importance to reorganize the integral process starting from the living environment of agricultural and livestock products, followed by production, processing, circulation, eating habits, intake and metabolism, and the disposal of the residues and wastes.

1.3.2 Data sources

To illustrate status of the key phrases within *Kanshokufuji*, large volumes of data were collected through various sources. Firstly, authorized data from the China Statistical Yearbook, China Agricultural Bulletin and data published through government websites, etc., were used to review the general situation and problems of the rural environment, agriculture and food. Secondly, other data published in academic monographs and journal articles was recited to provide additional proving or comparable data in analyses from general advances to the empirical findings within this book.

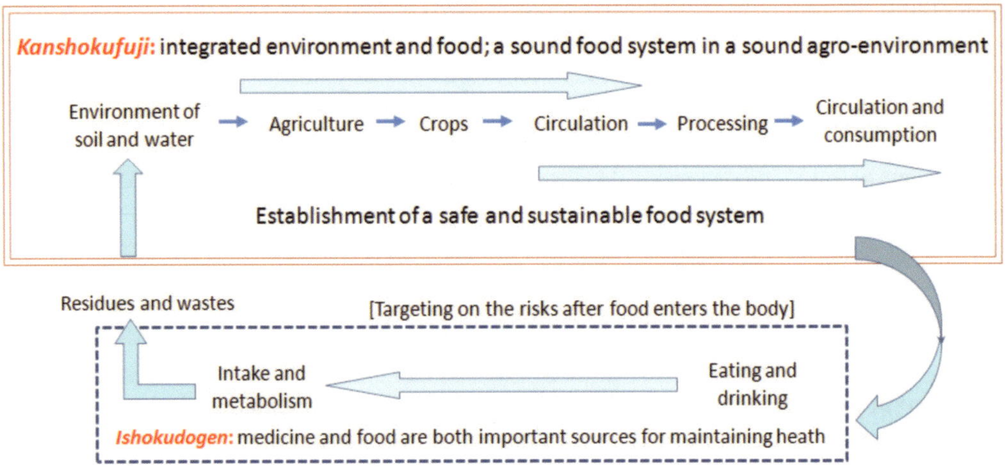

Source: Nanseki (2008) [23]

Figure 1.2 *Kanshokufuji* and *Ishokudogen*

In addition to the second-hand data mentioned above, much more first-hand data was collected through a variety of field surveys by the authors. The field surveys can be divided to two categories, both of which were conducted based on questionnaires and interviews. One category includes the rural and farmer surveys, while the other consists of consumer surveys. For instance, the general situation of the rural environment and agricultural production in different regions was studied through a survey of 21 villages in six provincial-level regions of eastern China. Using the data collected from a survey 560 household farms scattered across 21 villages of six provincial regions, we studied farmers' behaviours, perceptions and major affecting factors upon the application of fertilizers and pesticides, and farmers' confidence about the safety of their self-produced grain products. In light of the interviews of 168 sample dairy farmers of Inner Mongolia and Hebei Province, dairy farmers' perception of risk and their points of view on risk management strategies were studied. Based on a survey of 512 respondents from Beijing and Shanghai, we studied consumer perceptions on food safety and the major affecting factors. To examine consumers' attitude toward the traceability system, two interview surveys were conducted in Beijing and two samples with 209 and 214 respondents, respectively, were used to analyse consumers' risk awareness and willingness to pay for safety-certified food (Figure 1.3).

1.3.3 Major analysis methods

To fulfil the analyses mentioned above, a variety of methods are applied in the following chapters. Firstly, a series of empirical models are adopted to explore the implications behind the data collected. (1) Cobb-Douglas production function. From the perspectives of inputs change, institutional transition and technological progress, this book conducts a factor analysis of Chinese Agriculture Development after 1983. The macro-analysis is based on the time-series data issued by the government. After comparing the expressions of the two basic models of

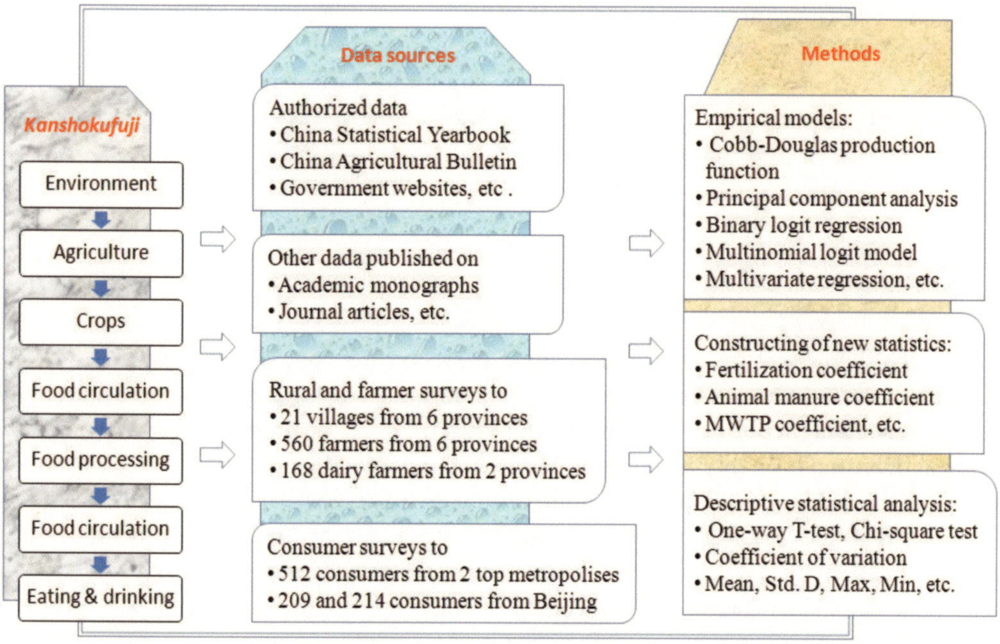

Figure 1.3 Theoretical framework of this book

Cobb-Douglas production function, the better performing model is adopted to demonstrate the relationships between economic growth and inputting factors. (2) Principal component analysis (PCA) is adopted to analyse the major factors affecting farmers' perceptions and strategies on food risk management. (3) Binary logit regression model is used to analyse the determinants of farmers' confidence on the safety of their self-produced grain products - farmers' behaviours in the application of agricultural chemicals. (4) Integrating with the choice modelling (CM) technique, the multinomial logit model is applied to examine consumers' attitude toward the traceability system. (5) Multivariate regression models are used to identify the significant determinants of pesticide application.

Meanwhile, further analyses are conducted through the construction or introduction of several new statistics. (1) In line with the soil properties represented by the geographical location in the National Fertilization Regionalization, the new statistic of *Fertilization coefficient* is formulated to isolate effects of farms' geographical location and planting structure, hence capturing farmers' propensities on fertilizing. (2) According to the *Animal manure coefficient* recommended by the National Environment Protection Bureau (NEPB), the total amount of animal manure and the major compositions of BOD_5, NH_3-N, TN and TP are calculated. (3) Through the *MWTP coefficient*, we managed to study consumers' marginal willingness to pay (MWTP) on the information provided by the traceability system, and to examine which factors affect consumers' willingness to pay for the traceability system.

In addition, descriptive statistical analysis methods, including the one-way T-test, Chi-square test, statistics of coefficient of variation, mean, std. D, max, min, etc., were widely used in this book to provide general scenarios or comparisons. In particular, the one-way T-test is used in identifying the major factors of consumer perceptions towards food safety. The analysis is conducted from the perspective of variables' significance in identifying the discrepancies among most of perceptions. Moreover, further analysis is conducted on the impact of demographic variables significant at the level of 0.01.

1.4 Organization of the book

Within the theoretical framework developed based on the conception of *Kanshokufuji*, the contents of the book are organized as shown in Figure 1.4 and Table 1.1. Chapter 2 studies the critical issues of agriculture, environment and food in China, including a factor analysis of the gross agricultural economy over recent decades, and a review of the food and environment. Based on the survey of 560 farmers within 21 villages of six eastern provincial-level regions, Chapters 3, 4 and 5 study the situation and determinants of rural environment, farmers' confidence about their self-produced agro-products, farmers' behaviours and perceptions and determinants of agro-chemicals' application, respectively. Chapter 6 analyses perceptions of dairy farmers on risk source and risk management, Chapter 7 studies consumer awareness and determinants in the top two Chinese metropolises upon food safety, while Chapter 8 aims to investigate consumers' risk awareness with regard to dairy products and their willingness to pay for certified safety food based on data from other surveys. As the concluding chapter, Chapter 9 includes an awareness comparison on environmental problems between farmers and consumers, followed by concluding and policy recommendations in light of the foregoing chapters.

Chapter 2: From the perspectives of inputs change, institutional transition and technological progress, this chapter conducted a factor analysis of Chinese agriculture development after 1983, when it began to develop as an independent industry by and large. This macro-analysis was based on the time-series data issued by the government, and the main model adopted was Cobb-Douglas production function. Through the application of SPSS, although several more independent variables were included into the model, the most effective factors eventually decided upon only contained the increment of chemical fertilizer, fixed agricultural assets, financial supports and the reduction of agricultural labour force, with the contribution rate of 53.70%, 15.57%, 4.77% and 1.66%, respectively. Furthermore, being the residual of the four variables above, it was calculated that technological progress contributed 24.30% to Chinese agriculture development in this period.

As a main conclusion, material inputs, including chemical fertilizer in the first place, composed the most important factor in agricultural development. As for the second factor, technical progress also promoted agricultural development to a considerable degree, while the contribution rate from institutional transition was comparatively low. Finally, a variety of suggestions were made on the topics such as secure application of chemical fertilizer, popularization of agro-technology, the increase of agro-capital, reduction of agro-labour, etc.

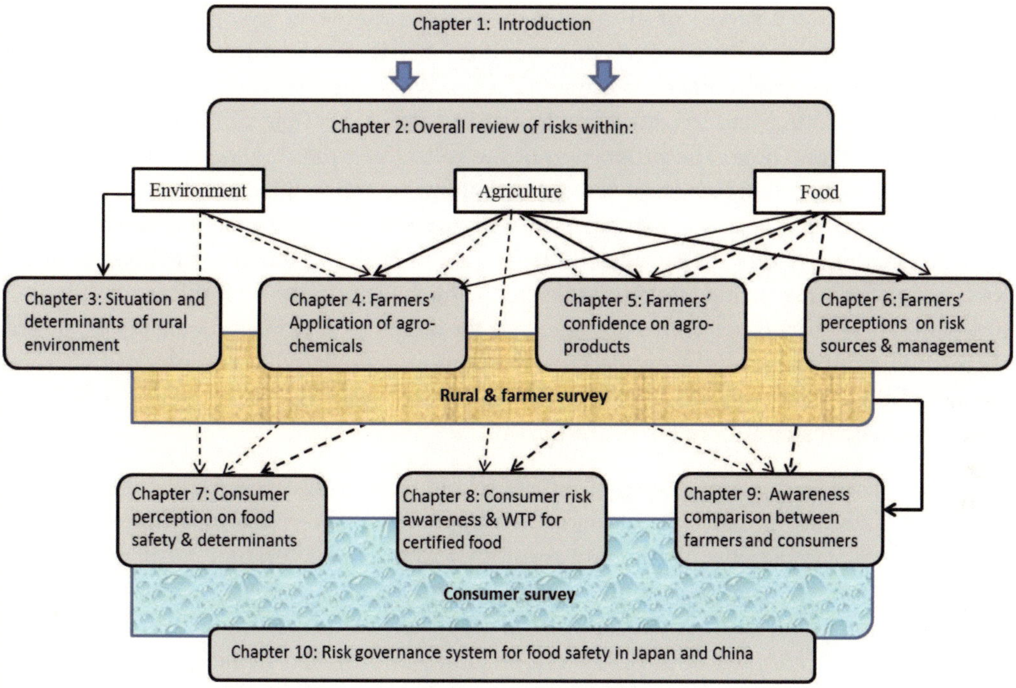

Figure 1.4 Organization of this book

	Environment	Food	Agriculture
Chapter 1	◎	○	○
Chapter 2	○	○	◎
Chapter 3	◎		○
Chapter 4	○	○	◎
Chapter 5		○	◎
Chapter 6		○	◎
Chapter 7	○	◎	○
Chapter 8		◎	○
Chapter 9	○	○	○
Chapter 10	○	◎	○

Note: ◎ refers to the main topics, while ○ denotes sub-topics of each chapter

Table 1.1 Topics of each chapter

Chapter 3: Based on a survey of 21 villages in six provincial-level regions of eastern China, this chapter studies that the situation of agricultural production in different regions and the relation of the rural environment to economic level. We found that 1and type, crop acreage, irrigation water source and irrigation methods have shown a certain difference between the north and south of China. The proportion of the collective rubbish and sewage disposal in rural areas is higher than industrial sewage and animal manure, the agricultural production and farmers' life guidance services are primarily supplied by government and agricultural extension centres. Within the surveyed villages, about 50% of them got subsidies from the government for constructing methane tanks, reducing the application of fertilizers, while only 10% of the surveyed villages were subsidized for the adoption of biodegradable plastic sheets. Meanwhile, empirical analysis revealed that the proportion of the collective rural rubbish and sewage disposal positively relate to the income level of farmers, while there is a negative correlation with distance from the nearest town.

Chapter 4: Based on the same survey of 560 household farms in six eastern provincial regions, this chapter studies farmer behaviours on the application of fertilizer, including the total amounts, main components of chemical fertilizer and the use of organic fertilizer. Then, it summarizes the farmers' perceptions, ranging from fertilizer choosing, field application, disposal of the used packages and awareness on the possible consequences of over fertilization. Nine indicators are adopted as the predictors, including information on the householders, land use and planting structure, household income and geographical location. The Fertilization coefficient is formulated to isolate the effects of farms' geographical location and planting structure, hence capture farmers' propensities on fertilizing. Through the adoption of binary logistic regression models, this chapter identifies significant determinants behind farmers' behaviours. As to the use of organic fertilizer, although demonstrated as statistically insignificant, possible impacts of chemical fertilization and breeding of livestock and poultry are included as predictors, in addition to the above indicators. Finally, a variety of policy recommendations are put forward, from increasing the fertilization efficiency of both chemical and organic fertilizer, to improving farmers' capability and awareness of scientific fertilization.

Meanwhile, this chapter studies farmers' application of pesticides, including the amounts of chemical pesticides, use of toxic pesticides and biological pest-control methods. Similarly, it summarizes the farmers' perceptions, ranging from choosing pesticides and field application to the awareness on the withdrawal period, possible consequences of overdosing and disposal of the containers. Thereafter, nine demographic indicators are incorporated as the candidate determinants, including information on the householders, land use and cropping structure, household income and geographical location. Through the adoption of multivariate OLS and logistic regression models, this chapter identifies significant determinants affecting farmers' behaviours. Finally, several policy recommendations are put forward, including the countermeasures to increase pesticidal efficiency, decreasing the use of toxic pesticides and improving farmers' capability and awareness on scientific application of pesticides.

Chapter 5: Following the publicity of a series of food safety incidents and the asymmetry of food information, consumers' confidence on Chinese food safety dropped dramatically. Approximately 70% of consumers are not confident of food safety. Compared with consumers, farmers, as managers and producers in agriculture production, have a lot of information of agricultural products. Based on a survey to 560 samples in six eastern provinces in China, this study selected 346 grain crop farmers to analyse farmers' confidence on their products. The data shows that more than 80% of farmers are confident on their products. In order to better understand farmers' confidence, we analyse influencing factors through the binary logit egression model. The result indicated use (or not) of manure, location and sowing area significantly affected farmers' confidence on their own agro-products. According to the result, some recommendations were proposed at the end of the chapter.

Chapter 6: The field survey was carried out in Inner Mongolia and Hebei Province in April and June 2010, respectively. A sample totalling 168 dairy farmers was available for analysis in this study. In this chapter, dairy farmers' perception of risk and their points of view on risk management strategies are studied. Risk is uncertainty that affects an individual's welfare, and is often associated with adversity and loss. In response to risky situations, farmers should be involved in risk management, making choices among alternatives so as to reduce the effects of the risks. The main research objectives are: to examine the dairy farmers' perception of risk and to examine the risk management strategies of dairy farmers.

Chapter 7: Based on a survey of 512 respondents from Beijing and Shanghai, this chapter studies consumer perceptions on food safety and the major affecting factors. In addition to the basic individual information of gender, age, educational background and employment, the demographic variables include professional experience of the respondent, family composition and also annual income. The perceptions consist of overall awareness about food safety, major sources of information and subjective reliability, understanding of the impact of environmental protection, main threats to food safety, the top sources of agro-pollution, most risky procedure or stage, viewpoints on the major responsibility bearer of agro-pollution and the best way to control agro-pollution. After the descriptive analysis on demographic characteristics and perception variables, the one-way T-test reveals that all the nine demographic variables are significant in identifying the discrepancies among most perceptions. Moreover, further analysis is conducted on the impact of demographic variables significant at the level of 0.01. Finally, a variety of policy recommendations are put forward, from strengthening the supervisory responsibility of the government, ensuring the all-round and effective supervision of food safety by the mass media and consolidating the supervision of key sectors, to accelerating the extension of environment-friendly technology.

Chapter 8: Consumers' risk awareness on dairy products and willingness to pay for certified safe food are studied based on related field surveys by the authors. In order to examine consumers' attitude toward the traceability system, an interview survey was conducted from September to October 2008 in Beijing, and 209 samples were collected in this survey. Data from another self-survey conducted in Beijing July 2008 by the authors is also used for analysis in this chapter. In this survey, 214 consumers were interviewed and applied as valid samples. The analysis of consumers' willingness to pay for certified safe food was carried out based on another survey. The analysis includes data of 209 respondents that correspond to 100% of the interviewed consumers in the survey site - Beijing.

Chapter 9: An awareness comparison on environmental problems between farmers and consumers was conducted in the first instance. Analysis of this section is based on two field surveys. The first one is the survey of farmers from six provincial regions as introduced in Chapter 3. While the other is a consumer survey held in Beijing, 2008. Consumers' attitude towards rice, vegetables, meat and milk were included in the questionnaire. There are 186 samples available from July and 209 samples available from September. The consumers' survey in July is mainly used in this case study. This survey includes respondents both from a supermarket survey and a home survey. In succession, based on the findings and conclusions from the foregoing chapters, a series of policy recommendations are put forward on risk management in China. Finally, perspectives on international cooperation between East Asia and the world are previewed from the standing of managing risks among food, environment and agriculture.

Chapter 10: As the last chapter, this chapter aims to establish an academic basis for the development of a risk governance system for food safety in East Asia for the cases of Japan and China. First, short histories of the food safety policy in Japan and China are reviewed. Secondly, the current statuses of the food traceability system in both countries are clarified. Thirdly, consumer perception on food safety is analysed from various perspectives. Fourthly, the current statuses of risk management at farm level (e.g., GAPs) in both countries are overviewed. Finally concluding remarks is given about further research. Various survey data, including several original surveys by the author, and government statistics are used for analysis in this chapter. The author's original survey on consumer awareness of food safety was done in Japan and China, 2008. The respondents of the preliminary surveys are 297 in total in China. The survey in Japan was an indoor group investigation using a survey slip. The survey in China was conducted by means of an individual interview. The author's original survey on consumer awareness of pork and milk traceability was conducted in China.

References

[1] China Ministry of Agriculture.China Statistical Yearbook 2012. Shenyang: Liaoning Education Press, 2012.5.

[2] Liu G.,et al. Current situation and measures of agricultural pollution in China. Studies in International Technology and Economy, 2006; 9(4):17-21.

[3] Liu G.,et al. Current situation and measures of agricultural pollution in China. Studies in International Technology and Economy, 2006; 9(4):17-21.

[4] CNSB (China National Statistical Bureau).Using of chemical fertilizers in different regions in China, http://www.stats.gov.cn/.

[5] Sun J. Review of agricultural pollution and preventive technology in China. Journal of Jishou University (Natural Science Edition), 2008; 29(5): 99-128.

[6] Hou B., Hou J., Wang Z. Farmers' perception on pesticide residue and its influence on pesticide application. Heilongjiang Agricultural Sciences, 2010(2):99-103.

[7] The survey shows that nearly 90% of Koreans distrust China's food safety, http://gb.cri.cn/27824/2012/08/16/6071s3812468.htm.

[8] Yang X., Li T. Analysis of agro-environment and agro-food in China. Food Safety 2003; (10).

[9] Liu S. Economic analysis about countryside environmental pollution in China, 2007.

[10] Lu Y., Xue H. Game analysis on agricultural non-point source pollution control in China. Agricultural System Sciences and Comprehensive Research, 2007, 23(3): 268-271.

[11] Shang Y. Game analysis on environment pollution control. Theory and Exploration 2005; 6:93-95.

[12] Hong D. Dual Control System and Environmental Problems in China. Journal of Renmin University of China 2000; 1(1):62-66.

[13] Zhu Q. The problems of peasants' environmental consciousness and the strategies to improve it. World Environment 2000; (4):24-26.

[14] Zhou J., Sun H. An investigation and analysis of environmental awareness of peasants in Jiangsu Province. China Rural Survey2009; (3):47-52.

[15] Yan X. Survey on rural environmental protection and villagers' environmental awareness - A case study in Xiangao countryside Nanjiang town PingjiangCounty of Hunan Province. New Rural Construction 2011; (1):31-32.

[16] Zhou J. Analysis on consumers' attitude, perception and purchasing behaviours on vegetables safety: basis on a survey in Zhejiang province. Chinese Rural Economy 2004; (11):44-52.

[17] Chern W., Rickertsen K., Tsuboi N., Fu T. Consumer acceptance and willingness to pay for genetically modified vegetable oil and salmon: A multiple-country assessment. AgBioForum 2002; 5(3): 105-112. From: http://www.agbioforum.org.

[18] Georges G., Rafia H. Consumers' Perception on Food Traceability in Europe. International Food &Agribusiness Management Association World Food & Agribusiness Symposium, Buenos Aires, Argentina 2006; (7):1-11.

[19] Wang Z. Perception on food safety and consumption decides: empirical analysis of individual consumers in Tianjin. Chinese Rural Economy 2003; 4:41-48.

[20] Zeng Y., Liu Y., Wang X. Analysis on willingness to pay for food safety with Hierarchical model: in the case of consumers' willingness to pay for moon cake additives. Journal of Agro-technical Economics, 2008; (1):84-90.

[21] Tonsor G. T. Consumer inferences of food safety and quality. European Review of Agricultural Economy 2011; 38 (2): 213-235.

[22] Li D., Nanseki T. Takeuchi S., Song M., Chen T., Zhou H. Consumer perceptions upon food safety and demographic determinants in China: empirical analysis based on a survey of 512 respondents. Journal of Faculty of Agriculture, Kyushu University, Japan 2012; 57 (2): 517-525.

[23] Nanseki T. Perspective of information technology for food safety. Japanese Journal of Agricultural Information Research 2008; 17 (4): 161-170 (in Japanese).

Chapter 2

Critical Issues of Food Safety and the Agro-Environment in China

Dongpo Li, Hui Zhou, Min Song and Teruaki Nanseki

2.1 Food safety in China

2.1.1 General situation

Agricultural pollution and environmental problems directly influence the food safety situation. Safe and healthy food only can be produced in a sound environment. In this chapter, agricultural pollution will be defined and environmental problems in China will be described.

With China's agro-production shifting to large scale and intensive operations, the huge amount of waste produced by the industry has not only worsened the circumstances where animals live, but also have adverse impacts on human health. The livestock industry has been urged to minimize pollutions caused by livestock production, explore effective waste management policies and technologies, and promote sustainable development of the industry.

2.1.2 Food safety issues

Food safety is a global issue nowadays. Food safety problems have caused many losses to consumers, producers and governments. Food safety problems have many causes such as economic problems, lack of technology and policy. In order to control the food hazard and food safety problems, the EU, Japan and the USA have been conducting research and have made great progress. According to the Chinese Food Safety Situation Report, the rate of certified food has been increasing since 2006 and the certified high-quality foods are becoming leading products in the markets. Chemical input residue, such as those from pesticides, chemical fertilizers, feed addictive, veterinary processes and others, is decreasing.

The current food safety situation is optimistic on the whole, but there are still some risks: new food safety problems may appear along with technology development. Checking several reports on food safety issues, we found that most food safety problems happened at either the production stage or at the processing stage. Additionally, the reasons could be: 1) chemical fertilizer or pesticide residue; 2) veterinary residue; 3) illegal addictive and 4) bacteria exceeding the standard.

Year	Issue	Reasons	Source
2010	Vegetable poisoned issue in Hainan Province	Pesticide residue	http://www.gov.cn/jrzg/2010-02/28/content_1543980.htm
2010	Vegetable poisoned issue in Guangxi Province	Pesticide residue	http://news.xinhuanet.com/legal/2010-04/01/c_1212625.htm
2010	Vegetable poisoned issue in Qingdao city	Pesticide residue	http://news.sina.com.cn/c/2010-04-10/033120043336.shtml
2009	Milk powder bacteria issues	Bacteria exceeding the standard	http://news.sina.com.cn/c/2009-02-02/043517130520.shtml
2008	Milk powder melamine issue	Illegal addictive	http://news.qq.com/zt/2008/juchong/
2008	Date poisoned issue	Illegal addictive	http://www.guyu.cn/content/48610.shtml
2006	Clenbuterol problems	Illegal feed and addictives	http://news.163.com/06/0916/04/2R45E8S70001124J.html

Source: internet survey

Table 2.1 Several food safety issues have occurred in China over recent years

2.2 Agricultural pollution and environmental problems

2.2.1 Definition of agricultural pollution

In the wider meaning, agricultural pollution is an ecological term that considers agricultural production as both the source and object of environmental deterioration. As shown in Figure 2.1, agricultural pollution is caused mainly by emissions of industrial wastes, including wasted water, gas and solid, urban-rural life sewage and rubbish, and pollution of agricultural production itself. Meanwhile, agricultural pollution can contaminate, farm land, water, etc., and result in quality risks for agricultural products [1].

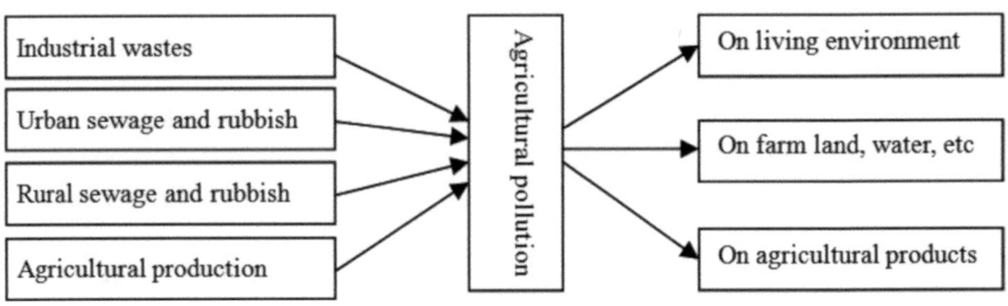

Figure 2.1 Mechanism of agricultural pollution in China

Compared with the other two sources, the non-point sources (NPSs) from rural sewage and rubbish and agricultural production come from scattered individual households and farms, which is difficult to control and thus of particular concern [2]. Simultaneously, they may result in serious environmental problems, including pollutions to soil and water due to over-application of chemical fertilizers and pesticides, pollution from wasted mulching film and other agricultural wastes, environmental pollution and deficiency of soil NPK (nitrogen, phosphorus and potassium) caused by burning straw, water contamination by livestock and poultry manure, etc. [3]. Moreover, it leads to over residues of nitrate, nitrite, heavy metals and even poisonous substances, resulting in direct impacts on the safety of agricultural products [4].

Agricultural pollution problems have happened often in China. There are 23 typical agricultural environmental problems which have occurred in China during 2000 to 2010. Agricultural environmental problems are mainly divided into three categories: agriculture is one of the sources which pollute the environment, agriculture is polluted by industry and other departments, and agro-resources exhausted by improper use and development of natural resources. More than half of the pollution problems are caused by agricultural itself. The main reasons are over use of pesticide, fertilizer, plastic sheeting, improper animal waste treatment and rural solid waste.

2.2.2 Chemical agro-input problems

Chemical fertilizer and pesticide residue is a big problem negatively affecting the agro-environment in China. Since 2000, China has become the biggest chemical fertilizer producer in the world. The rapid growth in China's per hector chemical fertilizer application has contributed significantly to the growth in grain production, but also caused many environment problems such as groundwater and underground water pollution. The improper use of chemical fertilizer and its residue is becoming one of the biggest problems in food safety - negatively affecting the environment [5-6].

Year	Chemical fertilizer	N: P: K	N	P	K	Compound fertilizer
2003	44.11	21.49	7.13	4.38	11.09	1:0.33:0.20
2004	46.36	22.21	7.36	4.67	12.04	1:0.33:0.21
2005	47.66	22.29	7.43	4.89	13.03	1:0.33:0.22
2006	49.27	22.62	7.69	5.09	13.85	1:0.34:0.22
2007	51.07	22.97	7.73	5.33	15.03	1:0.34:0.23
2008	52.39	23.02	7.8	5.45	16.08	1:0.34:0.24
2009	54.04	23.29	7.97	5.64	16.98	1:0.34:0.24

Source: China Statistical Press [8]

Note: the requested ratio by MOA among N, P and K is 1:0.37:0.25

Table 2.2 Use of chemical fertilizers in China (Unit: million *ton*)

According to the Ministry of Agriculture in China, the proper use ratio of chemical fertilizer among N, P and K is 1:0.37:0.25, but the reality is 1:0.34:0.24, which implies that the chemical use in China is not in proper balance, especially with regard to nitrogen overuse (Table 2.2). Based on research data from Henan Agricultural Bureau, only 1/3 of the chemical fertilizer is absorbed by plants, 1/3 gets into the air and 1/3 gets into the soil, and the chemical fertilizer pollutes both soil and air [7].

With the steady increase of agricultural production in China, extensively used pesticides have increased crop yields and produced high quality products over recent decades. Up to the end of 2010, the total amount of chemical pesticides produced in China amounted to 2.34 million tons, maintaining an average annual growth rate of 10.32% since 1985 [8]. China has become the largest producer, user and exporter of pesticides in the world. Meanwhile, the improper use of pesticides has become a major source of food safety incidents, which have resulted in serious threats (and losses) to the ecological environment, human health and economic development.

2.2.3 Animal manure waste problems

With the rapid development of the livestock industry, animal waste and hazardous residues of feed cause more and more pollutants. Some statistics suggest that in 2007 there were 439.895 million pigs, 105.948 million cattle, 285.657 million sheep and 9578.67 million poultry in China. According to the Animal Manure Coefficient recommended by the National Environment Protection Bureau (NEPB) (Table 2.3), we can calculate the total amount of animal manure at 2.147 billion tons, 42.445 million tons of BOD_5, 0.418 million tons of CODcr, 4.935 million tons of NH_3-N, 11.743 million tons of TN and 3.045 million tons of TP (Table 2.4), far greater amounts of solid waste from industry, which amounted to 1 billion tons in the same period of time.

Animal/poultry	Manure	Urea	BOD_5	CODcr	NH_3-N	TN	TP
Pig	398.00	656.70	25.98	26.61	2.07	4.51	1.70
Cattle	7300.00	3650.00	193.70	248.20	25.15	61.10	10.07
Sheep	950.00	--	2.70	4.40	0.57	2.28	0.45
Poultry	26.30	--	1.015	1.16	0.125	0.27	0.11

Source: China Statistical Press [8]

Note: poultry one is the average of chickens and ducks.

Table 2.3 Animal manure coefficient of NEPB (Unit: kg)

Animal/poultry	Manure	Urea	BOD_5	CODcr	NH_3-N	TN	TP
Pig	175.07	288.87	11.42	11.70	0.91	1.98	0.74
Cattle	773.43	386.71	20.52	26.29	2.66	6.47	1.06
Sheep	271.36	--	0.77	1.25	0.16	0.65	0.12
Poultry	251.92	--	9.72	11.15	1.19	2.63	1.10
Total	1471.80	675.59	42.44	50.41	4.93	11.74	3.04

Source: calculation based on the data of China Statistical Press [8]

Table 2.4 Animal waste discharged in 2007 (Unit: million *ton*)

2.2.4 Greenhouse gas emissions from the livestock sector

With rapid economic development, large amounts of greenhouse gases, such as carbon dioxide, CFCs, methane nitrous oxide and others, were produced. These gases in the atmosphere continue to accumulate and the average temperature of the mainland over the past century has increased significantly. The global average temperature increased 0.3-0.6 degrees, and the sea level rose 10-25 cm - indicating the fastest period of climate warming. After the Copenhagen Climate Change Conference, people began to pay more attention to greenhouse gases and climate change, and it became the most serious challenge facing the human race. The livestock sector is a major player in greenhouse gas emissions. According to the statistics from the Food and Agriculture Organization (FAO), the livestock sector is responsible for 18% of global greenhouse gas emissions measured in CO_2 equivalent.

Livestock production can result in methane emission from enteric fermentation and both CH_4 and nitrous oxide emissions from livestock manure management systems. Ruminants, such as cattle, are important sources of CH_4 because of their large population and high CH_4 emission rate as a result of their digestive system. Swine are also an important source of greenhouse gas, but less so than cattle. Poultry produce the least greenhouse gas. Animal waste will decompose in anaerobic conditions and produce methane, while the compost will produce large amounts of nitrous oxide.

	Province	CH_4 (10,000 ton)	%		Province	CH_4 (10,000 ton)	%
1	Sichuan	80.81	8.80	6	Xinjiang	50.15	5.46
2	Henan	77.42	8.43	7	Hebei	46.94	5.11
3	Inner Mongolia	70.45	7.67	8	Tibet	41.41	4.51
4	Shandong	55.55	6.05	9	Heilongjiang	39.15	4.26
5	Yunnan	52.41	5.71	10	Hunan	38.03	4.14
	Sub-Total	552.33	60.16				
	Total in China	918.17	100				

Source: calculated based on the data of China Statistical Press [8]

Note: 1. emission factors of livestock in sub-regions could not be available, so national emission factors are used to substitute.

Table 2.5 Top 10 provinces of methane emissions from the livestock sector

2.2.5 Veterinary residues in the livestock sector

The drug residues in livestock products contain drugs for animals and humans, disinfection chemicals, pesticides and other chemicals, the most prominent is residues of clenbuteral. Besides these, abuse of antibacterial drugs, hormones, vitamins and trace elements can cause residues of drugs. Drugs are not used properly, animals are slaughtered before the end of the drug-free period, or unapproved drugs are used as additives - all factors which can lead to contamination of the livestock produce. Once they are discharged with the manure, they will pollute the soil and water, and the environment. The potential threats of drug residues to human health include: allergy and abnormal response, taretogenicity, mutagenicity, carcinogenicity and bacterial infection. For example, dosages of hormones, such as cortisone and hydrocortisone which are widely used by veterinary surgeons, can cause few residues. But large scale over dosage can lead to large amounts of residues. However, their impacts on humans might be visible only after several generations, e.g., femininity in men and masculinity in women has some linkages with the abuse of hormones.

2.3 Development of Chinese agriculture and previous studies

2.3.1 Agricultural development over latest decades

At the end of 1978, China launched *Reforms and Opening-up,* thus breaking up the highly-planned economic institutions, and revitalizing the whole economy including those in rural areas. Up to the mid-1980s, as a prelude to the *Reform and Opening-up*, the *Household Contract Responsibility System* was expanded in rural areas across the country, where production teams were subsumed into 99% of villages and major production resources, symbolized by farmland, which were divided into household farms. After the reform, farmers could retain the rest of the goods and revenues as private property, once a certain amount of agricultural products or taxes were paid to the state as contracted. It released the long-term bound farming organizations and increased farmers' motivation with regard to agricultural productivity, hence agricultural development reached a high level within a few years. As a fundamental measurement, total grain yields amounted to 379 million tons in 1985, from 305 million tons in 1978. With this growth rate of some 25%, the problem of food security, which had puzzled China for a long time, was resolved by and large. In addition to benefiting national life and industrial development, it brought new opportunities to the overall economic reforms. At the same time, thanks to the effects of non-agricultural reforms, agriculture gradually developed as an industry capable of self-reliance.

By the mid-1980s, the rapid development of agriculture was realized primarily due to the powerful potential released by institutional reforms. By contrast, in subsequent periods, agriculture maintained its high-speed growth, under the progress of overall economic reforms. From 1983 to 2010, China's Gross Agricultural Output (GAO) rose from 275 billion yuan to 69319.76 billion yuan (current prices). Accounting for the influences of inflation, the GAO was 1242.70 billion yuan (using the constant prices of 1983), with an average annual growth rate of some 6.29 % being maintained in this period.

As shown below, factor analysis of China's agricultural development has been conducted in Lin et al. (1992) [9], Wang (2009) [10], Zhang (2008) [11] and other studies. However, these studies focused mainly on the specific changes of institutions, technology or inputs up to the early 21st century or even much earlier periods. That is, conducting comprehensive analysis and policy recommendations on the reasons behind agricultural development in recent periods remains a challenge to scholars when considering all the factors proposed above. Therefore, this chapter aims to clarify these issues, through factor analysis of China's agricultural development from the perspectives of inputs change, institutional transition and technological progress since 1983, when it began to develop as an independent industry by and large. Within this macro analysis on the national time-series data, the approaches adopted are mainly production functions.

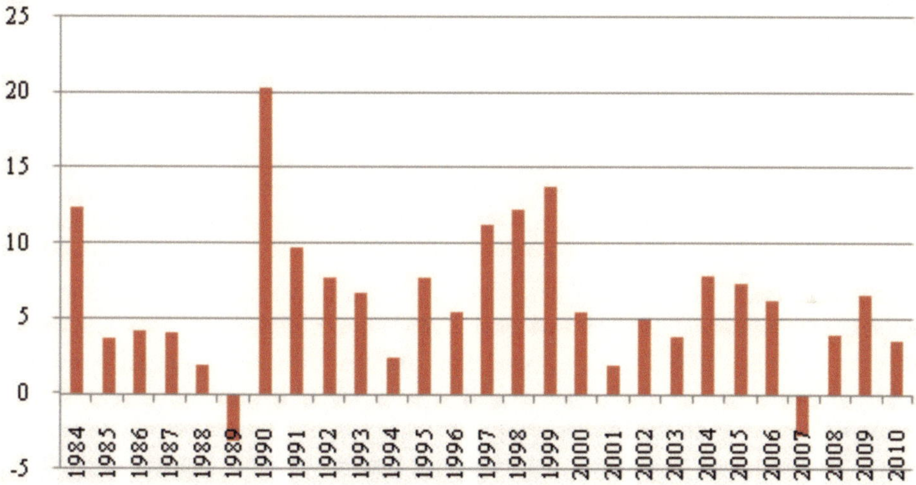

Source: China Statistical Press [8], China Ministry of Agriculture (2007) [12] (unit: %)

Figure 2.2 Annual growth rate of agricultural output in 1983-2010

2.3.2 Review of previous studies

In the study period, the Chinese government deemed rural areas as regions with great potential to expand domestic demands, with agriculture as the primary industry in the rapid and stable economic growth. Therefore, in order to stabilize the *Household Contract Responsibility System,* further reforms were conducted on the institutions of pricing agricultural products, agricultural taxation, etc. In addition, to increase agricultural productivity and farmers' incomes, more funds were inputted to the development of agricultural sciences and technology, especially the innovation and extension of advanced agricultural production resources and new breeds. Thus, modernization of agriculture has been promoted, thanks to these policies beneficial to agriculture and technological advances. However, agriculture did not develop continuously and at a fast speed, and significant differences still existed between the annual growth rates. In particular, after a minus growth rate of 4.78% in 1989, an upheaval of 21.83% was revealed in 1990 (Figure 2.2).

With regard to the zigzagged growth curve of GAO mixed with increases and decreases, many scholars have explored the causes from different perspectives. Lin (1992) [9] analysed the output elasticity of each factor in agricultural development from 1978 to 1984, using province-level panel data. According to the conclusion, as the most important factors in the first half of the period, rural economic institutional reforms from production teams to the *Household Contract Responsibility System* supported the increase of agricultural production. Meanwhile, the significance of institutional reforms diminished sharply in the latter half of the period. In succession, fertilizer application and technological progress (through the proxy variable of *T*) are measured as also contributing greatly. To analyse significant factors behind the development of Chinese agriculture, this study includes three areas of analysis: factor inputs, institutional changes and technological progress. However, as the study period is up to 1987, it necessary to conduct factor analysis of China's agricultural development within the following 20 years.

Huang et al (2005) [13] conducted an empirical analysis on the impact of changes in land ownership[1] on agricultural growth in the period 1949-1978, from the founding of the People's Republic of China to the *Reforms and Opening-up*. The conclusions of this study illustrate the different effects of each factor on gross outputs of agriculture at different stages of land ownership. This research has a long but demoded study period, and did not include the variable of technological progress. Aiming at understanding the impacts of the agricultural innovation system, Qiao et al (2006) [14] analysed the significant factors of Chinese agricultural development in the period of 1978-2004 (sub-divided into five periods), based on the model specified by Griliches (1963) [15]. However, insignificant variables were included in some models for different periods such as *labour* and *power* in 1978-1984 and 1996-2002. Meanwhile, the study periods were divided into so many stages, especially including a two-year stage of 2003-2004, this reduced the accuracy of statistical analyses with models of multivariate regression, etc., thus blocking the accurate measurement of the whole study period from 1983 to 2006. In addition, this study did not include the contribution of technological progress.

Based on the panel data of provincial-level regions, Zhang et al. (2008) [11] analysed the development of Chinese agriculture in the period 1949-2005. The result indicated that the physical inputs, particularly fertilizers and machinery, made a significant contribution to the total agricultural output, while farmland and labour contributed with lower or even negative ratios and large fluctuations. In this comprehensive empirical study, only the input elements were incorporated as determinants to agricultural development, while variables of technological progress and institutional change were excluded. In addition, Wang (2009) [10] studied the relationship between technological progress and economic development in agriculture, with an extended Cobb-Douglas production function. It concluded that agricultural development increased the funds inputted on agricultural technical progress, while the latter needs to feed

1 In the study period of this paper, land ownership in rural China passed through the stages of private ownership (1949-52), transition from private to collective ownership (1953-58), collective ownership through the people's commune (1959-62), and collective ownership of three subjects (people's commune, production brigade, production team), with the basis of production team (1963-78).

back the former mainly through the scientific conversion of concerning production elements and their organizations. Although this study was based on a long time period (1986-2004), the impacts of institutional change were not incorporated into the model (Table 2.6).

In previous studies, the development of Chinese agriculture was primarily attributed to three kinds of factors. The elemental inputs were the quantity of farmland, labour and agricultural assets, in addition to the liquid capitals of chemical fertilizers, etc. Institutional transitions referred to the changes of land ownership, agricultural price system, rural finance taxing forms, etc. Technology progresses included advances in farming methods related to increased production capacity of agricultural machinery and chemical fertilizers, and improved varieties of agricultural products. However, as noted above, there is still gap in research covering the period since 1983, when agriculture began to develop as an independent industry, with the adoption of the aforementioned factors to an integrated model, thus measuring the respective impacts on the development of Chinese agriculture. Meanwhile, further explorations are necessary in terms of the most appropriate indicator models to reflect the impacts from capital, land or other factors.

Therefore, with this in mind, based on data from the period 1983-2006 and production functions, and after thorough examination of the significance of each factor, this chapter selects a variety of indices with the availability of credible data, to demonstrate the impact of agricultural development. In detail, taking the 24-year period as a whole[2], the introduced time series data cover all the three types of variables as summarized above, i.e., inputs changes, technological progress and institutional transitions.

2.4 Production mode of agricultural development

2.4.1 Theoretical model of production function

In studies about sources of economic development, Cobb-Douglas production functions[3] are widely used to demonstrate the relationships between economic growth and inputting factors. As the original theoretical model, Cobb-Douglas production function is represented by the following formula:

$$Y = \beta_0 e^{\theta t} \prod_{n=1}^{N} x_n^{\beta_n} \tag{2-1}$$

2 To illustrate the impact of institutional changes, the author introduced several dummy variables and estimated the study period in different phases. However, the results did not show significant trends in terms of institutional changes, due to the short periods, thus illustrating the statistical insignificance of each model.

3 In the studies of relationships between economic growth and inputting factors, in addition to production function, cost function and profit function are often also used. Nevertheless, independent price variables are needed in both of the latter two functions. In addition, cross-sectional data were used in many prior studies on cost and profit functions [16]. In China, only part of the price data of production factors has been published. Therefore, this chapter conducts factor analysis of Chinese agricultural development with the adoption of production function.

Here, x_n represents the inputting factors of capital, labour, etc.; β_n is the elasticity of each factor; β_0 includes all the other factors as Total Factor Productivity (TFP); t is a proxy variable for the time trend variable of technological progress; β_0, β_n and θ are unknown parameters to be estimated.

Taking the natural logarithm on both sides of Eq.2-1, and calculating the partial differential of $\ln Y$ with t:

$$\frac{\partial \ln Y}{\partial t} = \theta \tag{2-2}$$

where θ represents the rate of technological progress [17]. Hence, the Cobb-Douglas specification of production function implicitly assumes the technological change effect is constant to the output Y [18]. Meanwhile, as the Cobb-Douglas specification of production function is homothetic, thus we assume that the substitute elasticity between different factors' constant to be 1[4] [16].

In theory, using the model described above, we can compute the contribution of technological progress over time. However, as the rate of technological change is not constant every year, there are extreme possible difficulties in observing the contribution over time. Therefore, another specification of production function is needed as follows:

$$Y = \beta_0 \prod_{n=1}^{N} x_n^{\beta_n} \tag{2-3}$$

Here, contribution of technological progress can be calculated with:

$$M_{Tech} = 1 - \sum_{n=1}^{N} M_n \tag{2-4}$$

That is, contribution of technological progress (M_{Tech}) is obtained as the residual of subtracting the contribution of other factors (M_n) from the growth rate of Y [19, 20]. This thus provides another basic method to estimate the contribution of technological progress, based on the Cobb-Douglas production function [21]. In this study, after comparing the results of the two models, the better performing Eq.2-4 is adopted.

4 Despite the single homogeneous assumption of substitutability between the elements, similar homogeneity (i.e., constant returns to scale) is not assumed for the returns to scale, which is determined by the parameters to be estimated. For example, if $\Sigma\beta n= 1$ means constant returns to scale; $\Sigma\beta n<1$ indicates the diminishing returns to scale; $\Sigma\beta n>1$ denotes increasing returns to scale.

	Period	Factor and production elasticity													R^2
		Land	Labour	Capital	Agro- power	Fertilizer	House-hold contract	Multi-crop	Cash crops	Price ratio	Price reform	Tax & costs reform	Public financial aid	Agro-tech	
Lin J. Y. (1992)	1978-84	-0.74	1.91	4.57		13.60	19.80	-0.82	1.56	-0.32	6.75			12.60	
	1984-87	-1.61	-2.95	1.88		2.26	0	0.88	1.17	5.36	-5.04			6.30	
Huang S., et al (2005)	1949-52	0.01	0.59		0.11	0.18					-0.25				0.850
	1953-58	0.50	0.42		-0.06	0.06					0.00				0.925
	1959-62	0.73	0.33		-0.09	0.04					0.28				0.913
	1963-78	0.34	0.40		0.03	0.09					0.24				0.816
Qiao Z., et al. (2006)	1978-84	-0.98	0.82		4.68	18.11	20.99			0.06					0.994
	1985-87	0.46	0.68		6.19	0.13				-0.52					0.998
	1989-95	-0.17	-0.84		5.88	2.05				-1.60		-0.56	11.80		0.998
	1996-02	0.15	0.14		1.46	6.96				0.06		-0.15	0.33		0.994
	2003-04	0.52	0.24		1.30	0.81				0.17		1.95	1.99		0.998
Zhang H., et al. (2008)	1984-87	-0.65	0.38		0.97	0.44									0.995
	1988-96	-1.54	0.94		2.86	0.23									0.917
	1997-03	-0.30	0.27		0.22	0.23									0.996
	2004-05	0.34	-0.49		1.05	0.64									0.994
Wang J. (2009)	1986-04		0.45			0.47								0.34	0.990

Note: the agro-power is the sum of the energy used in agriculture; price ratio is the ratio of price index of agricultural products and production resources; price reform is represented by ratio of product prices determined by the government; tax and cost reform is represented by the proportion of agricultural tax in total agricultural production; financial support refers to the proportion of fiscal inputs to agriculture in total public budgets

Table 2.6 Factor estimations of China's agricultural development in previous studies

2.4.2 Indicators and data

In this chapter, to describe the development of Chinese agriculture and the factors over the period, indicators shown in Table 2.7 are adopted. The data sources include Bulletin of Chinese Agricultural Development (2007) and China Statistical Yearbook (relevant years), published by China's Ministry of Agriculture and State Statistical Bureau. Considering the impacts of time trend, all the monetary values are calculated in the constant prices of 1983.

Var.	Description	Unit	1983	2006	Annual growth (%)
Y	Gross Agricultural Output (GAO)	billion yuan [a]	275.00	1242.70	6.78
Ld	Sowing area of agricultural plants	million ha	143.99	157.02	0.38
Ats	Value of fixed agricultural assets	billion yuan	53.40	237.82	6.71
Pw	Power of agricultural machineries	million kw	180.22	726.36	6.25
Fert	Amounts of chemical fertilizer	million ton	16.60	48.34	4.76
Lb	Number of agricultural labours	10000 person	316.45	294.05	-0.32
Rp	Price indices ratio of agro-products and inputting materials	%	101.36	99.70	-0.07
Rt	Ratio of agricultural taxes in fiscal revenue	%	4.25	0.95	-6.31
Rf	Ratio of fiscal agro-aiding funds in GAO	%	4.83	7.48	1.92

Note: [a] As the prime currency unit, 7.97 yuan = 1 US$ (middle exchange rate of 2006) and all the monetary values are calculated in the constant prices of 1983

Source: China Statistical Press [8] & China Ministry of Agriculture (2007) [12]

Table 2.7 Summary statistics of Chinese agricultural development in the period 1983-2006

In the first place, as the dependent variable, Gross Agricultural Output (Y) is the total output value of the final products of agricultural activities, including farming, forestry, animal husbandry and fishery. The gross output of each agricultural product (Y_t) is obtained by multiplying the price and physical volumes of production, and then converting to the constant prices of 1983.

Due to the existence of multiple cropping in agricultural production, Sowing Area of Agricultural Land (*Ld*), rather than the areas of arable land, is adopted [14]. In origin, Labour Force (*Lb*) should be represented with of total working days or hours in a year, etc. However, viewing from the perspective of the real status of Chinese farmers, it is difficult to accurately measure their labouring times. Meanwhile, relevant data is not found from the *China Statistical Yearbook*, *China Agricultural Yearbook*, and other sources. Therefore, referring to the earlier literature [9, 14], annual number of agricultural labours (10 thousand persons per year) is adopted in this study.

Agricultural capitals are divided into fixed and liquid capitals. The *value of fixed assets* (*Ats*) is the monetary expression of objects, tools and equipment directly used in agricultural production, borrowed or owned by farms over a relatively long period of several years. Power Agricultural Machineries (*Pw*) is the sum of energy with machineries used in ploughing, irrigation, harvesting and transportation, etc., within the agricultural activities of farming, forestry, animal husbandry and fishery. In order to identify appropriate variables to represent the fixed capitals, *Pw* and *Ats* are incorporated into the model simultaneously. At the same time, as the most important liquid capital, Amounts of Chemical Fertilizer (*Fert*) refers to the standardized quantity of nitrogen, phosphorus, potash and compound fertilizers used in agricultural production. Here, the standardization depends on the content of nitrogen, phosphorus pentoxide, potassium, etc., in different types of fertilizers.

Additionally, three indicators are included to reflect the impact and effectiveness of institutional reforms in the study period, concerning agricultural commodity prices, agricultural taxation, aids and assistance to agricultural production, etc. Price Indices Ratio of Agro-Products and Inputting Materials (price ratio, *Rp*) is the ratio of price index and producer price indices for agricultural materials in each year. The series of reforms carried out in the field of agriculture began from institutions of commodity prices in the early 1980s. Thereafter, the price system once generally controlled by the state is gradually being reformed over a long period. By 2004, fixed purchase prices were completely abolished and grain prices began to be fully determined by the market. Meanwhile, Ratio of Agricultural Taxes (*Rt*) is the percentage of agricultural taxes of national fiscal revenues in each year. The agro-supporting funds are mainly used to finance agricultural production, irrigation, climate forecasting, infrastructure, R&D, etc. The Ratio of Fiscal Agro-Supporting Funds (*Rf*) refers to the percentage of public funds within Gross Agricultural Output (GAO). Using these two indicators, we intend to evaluate the impact arising from the reforms of agricultural taxation and fiscal institutions. From 2000, rural tax reforms were included in the unified reforms managed by the central government and from 2006, agricultural taxes were fully abolished nationwide and subsidies supporting agricultural production began to be directly distributed to farmers. At the same time, reform of the budgetary expenditure on agriculture finance came into force. In 2004, to balance the socio-economic development of urban and rural areas, the No.1 document issued by the top authorities proposed the key guideline for the rural policies as 'Giving More, Taking Less and Loosening Control', stressing that the government would increase its input to rural areas and agriculture, and reduce taxes and fees collected from farmers. Meanwhile, in the same document, another policy agenda committed to transforming the lack of financial agriculture input[5].

2.4.3 Results of the estimation

In this study, factors analysis of agricultural is conducted through the development of an econometric model without the inclusion of time variable, based on the log-linear Cobb-Douglas production function as:

$$\ln Y = Constant + \alpha \ln Ld + \beta_1 \ln Ats + \beta_2 \ln Pw + \beta_3 \ln Fert + \gamma \ln Lb + \delta_1 \ln Rp + \delta_2 \ln Rt + \delta_3 \ln Rf + \varepsilon \quad (2\text{-}5)$$

where *Constant* is the intercept, α, β_i, γ and δ_i are unknown parameters to be estimated, and ε is the random item.

Although we can include all the above variables into the final model and obtain higher fitness, it is better to develop models by omitting redundant variables which hardly contribute the

5 In terms of the agricultural financial inputs, the total sum draws much more attention than the proportion of annual government expenditure.In recent years, the fiscal inputs to support agriculture have increased, while the proportion ofannual government expenditure decreased. In 1983-2006, the proportion decreased from 9.43% to 7.85% (China Statistical Yearbook).

total fitness. In econometric models, the change of determinant coefficient ΔR^2, the change of $F(\Delta F)$ and the probability significance of $p_{\Delta F}$ are referential in selecting the variables [22]. After removing the insignificant variables according to the probability significance of $p_{\Delta F}$ obtained by the software SPSS, the combination of the explanatory variables in the final model include four significant variables as shown in Table 2.8.

All the significant F and t-test at the level of 5%, the Adj.R^2 of 0.99 and Durbin-Watson value of 2.285 indicate good statistical fitness. In addition, fixed assets, chemical fertilizers and ratio of fiscal agro-supporting funds are all estimated with positive elasticity. Although the negative elasticity of agricultural labour is adverse to the general economic assumption, it meets with existence of surplus amounts of labour in Chinese agricultural production. In previous studies, both Lin (1992) [9] and Zhang et al (2008) [11] measured the negative elasticity of labour productivity. Therefore, this model estimates well China's agricultural and economic growth in the study period.

Variable	**Constant**	**Lb**	**Ats**	**Fert**	**Rf**
Elasticity	7.672**	-0.824**	0.159***	0.988***	0.306***
t	(2.436)	(-2.665)	(2.848)	(12.265)	(4.824)
Indicator	Sample size	F	Adj.R^2	D-W	
Value	24	586.085***	0.99	2.285	

Note: ***, **and * represent statistical significance in the level of 1%, 5% and 10% respectively

Software: SPSS 13.0

Table 2.8 Estimation of the production elasticity

With regard to the causes of the significant ratio of fiscal agro-supporting funds and insignificant pricing factors, these may be due to lower prices of agricultural products compared with the prices of fertilizers and other inputting industrial products, thus farmers find it difficult to attain positive agricultural productivity. In terms of the sowing area of agricultural plants, insignificance may result mainly from multicollinearity, as high relation coefficients of 0.95 and 0.87 exist between this variable and the value of fixed assets and fertilizer, respectively. Meanwhile, viewing from changes of inputs over the study period, when sowing area of agricultural plants increased 9.05%, the value of fixed assets and amount of fertilizer increased 345.38% and 191.26%, respectively. With respect to the major crops, acreage of grains and cotton declined 7.50% and 11%, respectively, thanks to the increased per unit yields of 38.88% and 63.45% - total yields eventually increased by 28.45% and 45.48%, respectively. Similarly, total yields of oil crops increased by 189.99%, due to the increased per unit yields of 77.12%. To sum up, in the study period, comparing with the physical inputs of fertilizers and fixed assets, etc., hence the increased in yield per unit, sowing area exerted slightly smaller effects, thus the insignificant result in the quantitative model is plausible.

2.4.4 Contribution of each factor

In the study period, although the gross agricultural output increased by 351.89%, the amount of agricultural labour decreased by 7.08%, from 316 million to 294 million. In addition, the value of agricultural fixed assets increased from 53.4 billion yuan to 237.8 billion yuan, increasing more than four times; the three-fold increase of fertilizers meant an increase of 48.34 million tons (rising by 16.60 million tons). The ratio of fiscal agro-supporting funds in gross agricultural output also rose from 4.83% to 7.48%. Based on the multiplication of these changes on each factor and the corresponding elasticity, the contribution rate of agricultural growth can be assessed for each factor using percentage within total agricultural output. Furthermore, as shown in Eq.2-4, contribution of agricultural technological progress can be estimated by subtracting the contributions of the other factors (Table 2.9). In addition, as investment on agricultural R&D is already included in financial support for agriculture, the investment on agricultural R&D from the government is not included in the technological progress in this context.

		Y	Lb	Ats	Fert	Rf	Tech [a]
Total change (%)	(1)	351.89					
	(2)		-7.08	345.38	191.26	54.87	—
Elasticity	(3)		-0.82	0.16	0.99	0.31	—
Contribution (%)	(3)×(2)/(1)		1.66	15.57	53.70	4.77	24.30

[a] Contribution of technological progress (*Tech*) is calculated based on Eq.2-4

Software: Excel 2007

Table 2.9 Contribution of each factor (1983-2006)

According to the results of Table 2.9, within the growth rate of 351.89% of gross agricultural production in the study period, the increased amount of fertilizers, value of fixed assets, amount of financial support and the reduction of agricultural labour force, contributed 53.70%, 15.57%, 4.77% and 1.66%, respectively. Meanwhile, being the residual of the four variables, technological progress contributed 24.30% to Chinese agriculture development. Among the factors, increase in the amount of fertilizer inputs is the most significant factor, followed by technological progress. These two factors accounted for 78% of gross agricultural output growth, constituting major causes of Chinese agricultural development over the study period.

In succession, the value of fixed assets indicates great increases but small elasticity, thus the contribution remained 15.57%. As a proxy of institutional changes, the ratio of fiscal agro-supporting funds in GAO makes a small contribution of 4.77% in the study period. In terms of the minus and small elasticity of agro-labour, 1.66% is contributed due to the decreased numbers over the study period.

These results are in line with the conclusions of prior studies. Firstly, with regard to the basic agricultural production resources, the detected significant effects of chemical fertilizer are similar to those found in Lin (1992) [9], Zhang et al (2008) [11] and Qiao (2006) [14] and many other studies. As a key factor in second place, the importance of technological progress is measured in Wang (2009) [10] and other prior literature. As for the negative elasticity of agro-labour, which is in line with Lin (1992) [9] and Qiao (2006) [14], this indicates that transferring of agro-labour numbers have contributed to Chinese agricultural development. The major reasons behind this include engaging in other sectors enabled the farmers to obtain more funds to invest in fertilizers and fixed agricultural assets. Meanwhile, the non-agricultural experiences are beneficial in improving farm management and the trade of agro-products.

2.4.5 Major conclusions and recommendations

1. Findings and conclusions

In this chapter, a factor analysis of Chinese agriculture development in the period 1983-2006 is conducted, from the perspectives of inputs change, institutional transition and technological progress. As a result, we did not ascertain new findings and similar results to prior studies were obtained, using comprehensive perspectives, overall and long-term modelling and a comparison of different models in measuring the contribution of technological advances, etc.

With regard to the statistical significance of each factor, with the increment of chemical fertilizer was in first place, fixed agricultural assets next, followed by financial support and the reduction of agricultural labour force – these all constituted the major factors supporting China's agricultural development in the study period. In previous literature, different factors were detected as the first factor in different stages, such as agricultural technology in Lin (1992) [9], agricultural machinery and financial assistance in Qiao (2006) [14], agricultural machinery and labour force in Zhang et al (2008) [11], etc. (Table 2.6). In contrast, increased input of fertilizer is measured as the most important factor for China's agricultural development in the period 1983-2006, with an overwhelming contribution share. In addition, as the second factor, technological progress is concluded as supporting agricultural development with a considerable share of contribution. Different from the models in Zhang et al (2008) [11] and Qiao (2006) [14], Wang J. (2009) [10] considered the significance of agricultural technology, although it was measured as contributing the lowest share among three types of factors. Inaccurate measurement of the contribution of agricultural technological progress will inevitably lead to a misunderstanding of the significant factors and thus policy recommendations with regard to agricultural development. Finally, although Lin J. Y. (1992) [9] and Qiao Z., et al (2006) [14] have suggested that institutional changes is a key factor for China's agricultural growth since the middle of the 1980s, this study shows that compared with the other factors, institutional changes holds a relatively low contribution share. It suggests that since the mid-1980s, despite a series of policies in favour of agriculture, few fundamental institutional changes, like the household contracting system, were carried out or institutions were not implemented effectively.

2. Policy recommendations

Based on the above conclusions, the following policy recommendations can be made to accelerate agricultural development. First, the increased input of chemical fertilizers has contributed significantly and the amount is expected to keep increasing in the future. Nevertheless, the realization of sustainable and environmentally friendly agriculture has become an important issue. Therefore, for the safe application of chemical fertilizers, the government needs to extend soil surveying techniques, and promote the proper classification and appropriate amounts of fertilizers. Thereby, increase agricultural productivity while savings fertilizer costs and protecting the environment.

Meanwhile, as another important factor, advances in agricultural technology should be accelerated from the three perspectives of R&D, extension and funds. At present, although agricultural technology has developed rapidly in China, problems still remain in transferring and spreading the techniques to the fields. When analysing the causes, this can be attributed to the overwhelming ratio of household management, thus the small sizes of farmland in agriculture. Individual farms are limited in willingness and ability to introduce new agricultural technologies. In the governmental institutions specializing in extending agricultural technologies, there is poor connection between staff incomes and their achievements in extending agricultural technology, thus there is a lack of initiative and this can be pointed to as another reason. Thus, in addition to enlarging the managerial scales through encouraging the transfer of farmland use rights, the adoption of more marketing mechanisms to the agricultural extension institutions at the grassroots level, simultaneously constitutes an urgent measure.

At the same time, countermeasures are needed to serve the needs of agricultural labour, the number of which is estimated with a significant but minus production elasticity as shown above. To reduce the amount of agricultural labour, further endeavours are necessary to strengthen the non-agricultural vocational training of rural labour through the *Sunshine Project*[6], accelerating the reform of the family registration system, so as to shift surplus labour to both the urban areas and local non-agricultural sectors.

In addition, as an important factor of institutional change, fiscal agro-supporting funds need to be increased. Despite the relatively low capital elasticity, the increase of fiscal agro-supporting funds is highly beneficial to increase the value of fixed assets and the promotion of agricultural technology advances. Thus, to strengthen the reforms of the financial budget on agriculture and transform the lack of agricultural fiscal inputs, it necessary to ensure sources of funds are channelled to agriculture, from a series of sources including governmental departments and financial institutions. Specifically, the main roles of government include the investment in agricultural infrastructure construction, and subsidies on the purchase of agricultural machinery and good seeds. Meanwhile, financial institutions are expected to create preferential prerequisites to provide more loans to farmers.

3. Further discussion

In recent years, with China's rapid economic development, agricultural inputs are being increased, and the extension of agricultural technology is being enhanced. In addition, with

---

6 *The Sunlight Project is a series of technical and vocational training programmes targeting the rural labour, carried out by the Chinese government since 2004. The project aims to improve the quality and skills of rural labours, thus promoting their employment in rural non-agricultural sectors and urban areas.*

the overall abolition of agricultural taxes and direct aid to agricultural production, etc., a series of agro-supporting policies were carried out by 2006. Therefore, although this study could not fully grasp the impact of these policies, further studies are necessary to assess China's overall agricultural growth factors, using annual data since 2007.

References

[1] Zhu L., Liang S., et al. Investigation on the agricultural pollution and prevention modes in different regions, by Institute of Agricultural Economics and Development, Chinese Academy of Agricultural Sciences, August, 2009.

[2] Edwin D. O., Zhang X., et al. Current status of agricultural and rural non-point source pollution assessment in China, Environmental Pollution 2010; (158): 1159–1168.

[3] Jia R., Lu Q., et al. Current Situation, causes and countermeasures of the agricultural pollution in China, Chinese Journal of Review of China Agricultural Science and Technology2006; 8(1): 59-63.

[4] Lin H., Li X. Effects and countermeasures of agricultural pollution to the quality safety of agricultural products, Chinese Journal of Ecological Economy 2009; (9): 146-149, 153.

[5] Wu S. 1991, Preliminary investigation of water N concentration within rural area in lower reach of Yangtze in China. Paper presented at the International conference on Agriculture and environment. Ohio State University, Columbus, Ohio, 10-13 November, 1991; 53(2): 569-574.

[6] Liu M., Du L., Zhang X., Farmers' willingness on organic fertilizer application based on logit model and influencing factors Journal of Anhui Agricultural science 2010; 38 (9): 4827-4829.

[7] Wang Z. Analysis on the influencing factors of fertilization behavior of farmers: Also on scientific fertilization and environmental pollution. Annual Report of Economic and Technological Development in Agriculture. China Agricultural Press 2011: 211-219.

[8] CSP (China Statistical Press). China Statistical Yearbook: http://www.stats.gov.cn/tjsj/ndsj/.

[9] Lin J. Y. Rural Reforms and Agricultural Growth in China, the American Economic Review 1992; 82 (1): 34-51.

[10] Wang J. Study on the relationship between inputs on R&D and economic growth in Chinese Agriculture, Chinese Journal of Journal of Agritechnical Economics 2009; (1): 103 -109.

[11] Zhang H., Chen Z. Analysis on the factor contribution to Chinese agriculture: study based on unstable panel data model, Chinese Journal South Economy 2008; (1): 61-75.

[12] CMA (China Ministry of Agriculture). Bulletin of Chinese Agricultural Development 2007: http://www.agri.gov.cn/sjzl/baipshu.htm.

[13] Huang S., Sun S., Gong M. Impact of Land Ownership Structure on Agricultural Economic Growth: An Empirical Analysis on Agricultural Production Efficiency in Chinese Mainland (1949—1978), Chinese Journal of China Social Sciences 2005; 3: 38-47.

[14] Qiao Z., Jiao F., Li N. Change of rural institution and agricultural economic growth: an empirical analysis on agricultural economic growth in China 1978-2004, Chinese Journal of Economic Studies2006; 7: 73-82.

[15] Griliches Z. The Source of Measured Productivity Growth: United States Agriculture, 1940-1960, Journal of Political Economy 1963; 71 (4):311-346.

[16] Kuroda Y. Study on Technological Changes in Japanese Agriculture: an outlook, included in Izumida, Yoichi: 50 years of modern economic analysis of agriculture and rural areas, Press of Agriculture and Forestry Statistics Society; 2005: 127: 133.

[17] Sakano S., Kuroda S., Suzuki Y., Minotani T. Economics Quantitative Analysis Series, Vol. 4, A lied Econometrics, Taga Press; 2004: 279-281.

[18] Coelli T. J., Prasada Rao D.S., et al. An Introduction to Efficiency and Productivity Analysis (2nd edition), Springer; 2005: 211-213.

[19] Ogawa K., Tokutsu I., 2002, Guidance on the empirical analysis of Japanese Economy, Yuhikaku Publishing Co., Ltd: 261.

[20] Sakano S., Kuroda S., Suzuki Y., Minotani T., 2004, Economics Quantitative Analysis Series, Vol. 4, A lied Econometrics, Taga Press: 279-281.

[21] Inamoto S. Measurement of technical progress and aggregate production functions in Japanese agriculture: an outlook, The Farm Accounting Studies, Research Information Repository, Kyoto University1969; 37 (3): 80-91.

[22] Murase Y., Takada H. Multivariate Analysis with SPSS, Tokyo: Ohmsha Press; 2007: 170-171, 203.

Chapter 3

Situation and Determinants of Agro-Environment

Tinggui Chen and Min Song

3.1 Introduction

Along with China's fast economic development, environmental problems in rural areas are becoming increasingly severe. To tackle this problem in the rural environment, the Chinese government proposed the building of a 'Socialist New Countryside' since 2005, promoting the capacity and management to create clean villages as one target in this nationwide project. In April 2005, rural rubbish began to be included in the revised version of the Solid Wastes Pollution Prevention Law. "Advice on strengthening the rural environmental protection", released by the State Council on November 13, 2007, put forward specific targets to improve the quality of rural drinking water, and control soil and agricultural pollution sources. Furthermore, "Implementation of awarding for promoting the governance and settlement of rural environment outstanding issues", released by State Council on February 27, 2009, raised the objectives and effectiveness requirements of rural environment governance. Now Beijing, Shanghai, Zhejiang and other developed areas are implementing rubbish treatment projects and the awareness of environmental protection among rural residents is becoming stronger and stronger [1-2].

Based on a nationwide survey of 141 villages, Tang et al (2008) [3] analyses the problem of environmental pollution in rural China. The study shows that 49.53% of the villagers think civil rubbish is the main source resulting in environmental pollution in rural areas, while 32.71% of the villagers attribute the most important sources to industrial pollution, and in the third place, 9.35% of them selected chemical fertilizers as the major source of environmental pollution in rural areas. In order to have a better understanding of the current situation of rural environmental pollution and its major determinants, Huang et al (2010) [4] conducts a nationwide survey, with a representative sample of 101 villages from five provinces. The result indicates that about 44% of the villagers think that their environment became worse over the past 10 years. The following econometric analysis shows that rural enterprise development and local township expansion (or rising population density) has negatively affected the rural environment, while government's efforts have a positive effect on slowing down the pace of environmental pollution.

Based on the survey of 21 villages of eastern China's six provincial regions, this chapter aims to analyze the situation of agricultural production, the public service in rural regions and the relationship between the rural environment and economic level.

3.2 Samples of the survey

To understand the present situation and farmers' perceptions on agricultural pollution, we conducted the survey with questionnaire-based personal interviews to collect first-hand data. In the first section, our questionnaire contains the basic characteristics of each village, including general economic situation, land and land use. In the second section of the questionnaire, we investigate the situation of agricultural production, including crops, animal products, aquatic products, certificated agricultural production and irrigation. The following questions are designed to ascertain information about the rural environment, including rubbish collection, disposal of liquid and solid waste, construction of methane tanks, etc. In the final section, we collect rural public services data, such as agricultural production, rural life, government subsidies and village public expenditure. Simultaneously, we designed another questionnaire to understand farmers' perceptions on agricultural pollution, including basic characteristics of each farm, disposal of household rubbish, farmers' selection and application of fertilizers, pesticides and veterinary drugs, farmers' perceptions on agricultural pollution, and information about and recognition of safe agricultural production.

In January to March, 2011, we surveyed 21 villages and their 560 farms in eastern China's six provincial regions, including Beijing, Hebei, Shandong, Shanghai, Jiangsu and Zhejiang (Figure 3.1). The sampled area covers three major gain growing provincial regions and rural regions affiliating to the top two metropolises of China. The former three regions represent the northern mode of agricultural production in the Yellow River Basin, while the latter three demonstrate the characteristics of agricultural production in south China's Yangtze River Basin. With regard to topographic types, farms located on plains, hills and mountainous regions, and villages in inland areas, seaside areas and areas adjoining the metropolises are sampled. In addition to the staple grain crops of wheat, rice and corn, the other major agricultural products, including cotton, vegetables, fruit, oil crops, etc., and the main livestock, poultry, aquaculture products are being grown and cultivated in the sampled farms.

3.3 Basic situation of the sampled areas

In six sampled areas, the GDP of Jiangsu Province in 2010 is the largest, accounting for 10.33% of the national share. Beijing's GDP is the smallest, accounting for 3.52% of the national value. Being respectively the capital and the largest city in China, Beijing and Shanghai serve as central metropolises of the economy, and hence their GDP per capita is higher than in other provinces. Shanghai's GDP per capita amounts to 74,548 yuan, being the highest in the six areas, followed by Beijing with 71,938 yuan per capita. Hebei Province's is the lowest, less than the national

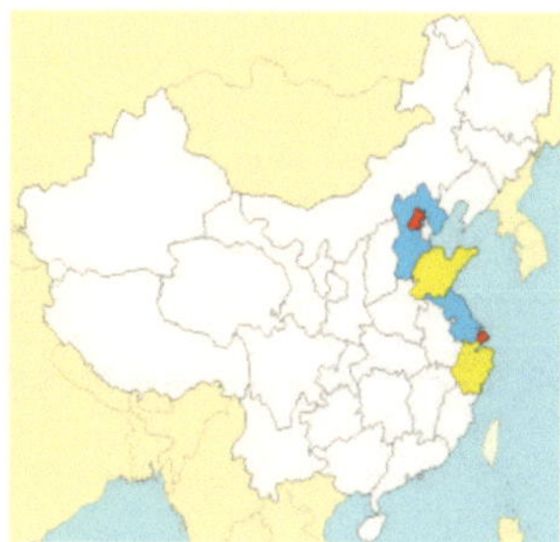

Source: revised based on http://www.chinamapxl.com/

Figure 3.1 Location of the sampled areas

average value and only accounts for 2/5 of that in Shanghai. Overall, the economic level of the three areas in the south is higher than in the three northern areas. Except for Hebei, the sampled areas possess significantly higher GDP per capita than the national average (Table3.1).

The largest agricultural output value in 2010 is from Shandong Province, amounting to 367 billion yuan, and accounting for 9.93% of the country's agricultural output value. Meanwhile, its arable land area is the largest, accounting for 6.17% of the total area of the country. The agricultural output value is almost the same in Beijing and Shanghai, accounted for only 0.42% of the national value. Their arable land areas are also very close, 0.19% and 0.20% of the country, respectively. When calculating the output per hectare, we find that Beijing is the highest, followed by Shanghai, with the lowest being Hebei, lower than the national average value.

Analyzing the areas of arable land per capita, the values of the six provincial areas are lower than the national average. Of these, Shanghai possesses the smallest arable land area per capita of only 0.28 mu, far less than the national average of over 2 *mu* per capita. By contrast, Hebei Province has the largest value of 1.98 mu, being close to the national average. Moreover, comparing values of the six provincial regions, we can conclude that the resource endowments of arable land per capita of the three northern regions, i.e., Beijing, Hebei and Shandong, are better than that of the three southern regions of Shanghai, Jiangsu and Zhejiang (Figure 3.2).

	GDP		Total population (end of year)		GDP per capita	Total agricultural output		Arable land area	
	Trillion yuan	%	Million person	%	1000 yuan	Trillion yuan	%	Million ha	%
China	40.12	100.00	1340.91	100.00	29.92	3.69	100.00	121.72	100.00
Beijing	1.41	3.52	19.62	1.46	71.94	0.02	0.42	0.23	0.19
Hebei	2.04	5.08	71.94	5.36	28.35	0.25	6.69	6.32	5.19
Shandong	3.92	9.76	95.88	7.15	40.85	0.37	9.93	7.52	6.17
Shanghai	1.72	4.28	23.03	1.72	74.55	0.02	0.42	0.24	0.20
Jiangsu	4.14	10.33	78.69	5.87	52.64	0.23	6.14	4.76	3.91
Zhejiang	2.77	6.91	54.47	4.06	50.90	0.10	2.82	1.92	1.58

Note: 100 dollar=634 yuan, http://www.boc.cn/sourcedb/whpj/, 2012.9.21

Source: China Statistical Press (2011) [5]

Table 3.1 Basic information of the surveyed areas in 2010

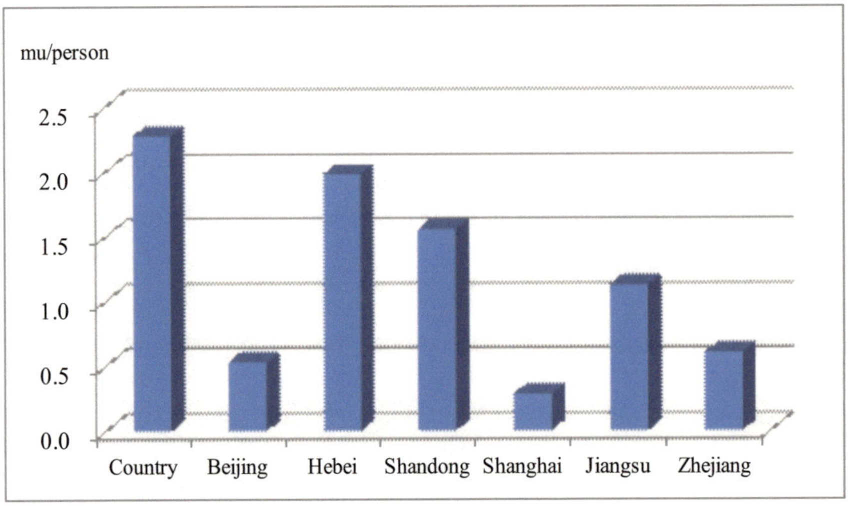

Note: 1 ha=15 mu.
Source: China Statistical Press (2011) [5]

Figure 3.2 Arable farmland area per capita in 2008

Per capita income of rural households and per capita GAP in the six regions show the same structure. Per capita income of all the areas is higher than the national average. Shanghai and Beijing's are significantly higher than the other four areas. Overall, the incomes of the southern areas are higher than that of the northern areas (Figure 3.3).

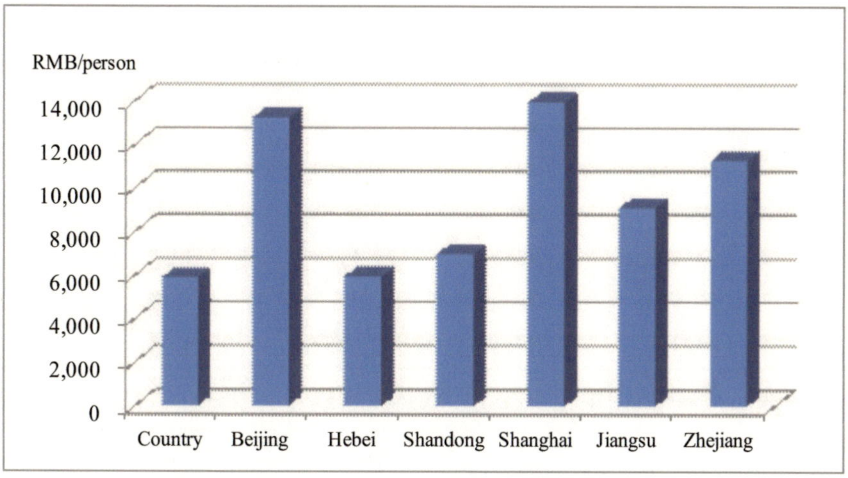

Note: 100 dollar=634 yuan, http://www.boc.cn/sourcedb/whpj/, 2012.9.21.
Source: China Statistical Press (2011) [5]

Figure 3.3 Net income of rural household in 2010

The irrigation area proportions of the total arable land area in the six regions are all higher than the national average. Among them, the highest proportion is Beijing reaching 91.25%, 1.8 times the national average. Shanghai's is 82.39%, the second highest, and the lowest is in Shandong Province, being 65.94%.

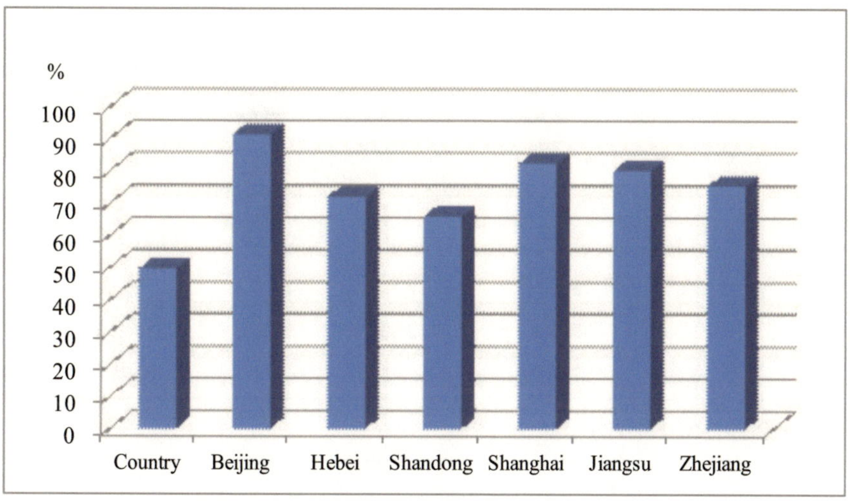

Source: China Statistical Press (2011) [5]

Figure 3.4 Proportion of irrigated arable farmland in 2010

	Total	N	P	K	Compound fertilizer
China	456.94	193.38	66.19	48.18	147.76
Beijing	589.99	296.50	37.98	31.07	224.43
Hebei	511.08	242.30	74.89	42.48	151.40
Shandong	632.47	216.39	66.37	61.74	287.97
Shanghai	485.33	253.73	41.40	24.18	166.01
Jiangsu	716.05	376.86	100.19	43.66	195.33
Zhejiang	479.99	273.32	62.32	37.33	107.04

Source: China Statistical Press (2011) [5]

Table 3.2 Amount of chemical fertilizers used in 2010

The amounts of chemical fertilizer application in the six surveyed areas are all higher than the national average. The highest is in Jiangsu Province, up to 716 kg per hectare, much higher than the national average of 457 kg, 632 kg in Shandong Province followed. Shanghai and Zhejiang are almost the same. Specifically, the amounts of nitrogen fertilizer application in the six surveyed areas are all higher than the national average, but the case in relation to phosphate, potash and compound fertilizer varies. The amount of phosphorus fertilizer application in Beijing and Shanghai is less than the national average. The amount of potassium fertilizer application in only Shandong is above the national average. The amount of compound fertilizer application in only Zhejiang is below the national average. The differences of fertilizer application may be mainly due to crop differences and soil differences.

	Rice	Wheat	Maize	Cotton	Peanut	Rapeseed
China	6553.03	4748.44	5453.68	1229.42	3455.45	1775.10
Beijing	6333.33	4609.51	5620.50	1150.00	2990.46	440.00
Hebei	6805.12	5084.51	5014.71	979.26	3517.36	1311.57
Shandong	8294.34	5779.55	6537.68	944.85	4211.78	2792.13
Shanghai	8327.65	3896.76	6659.14	1453.72	2761.80	2195.15
Jiangsu	8091.90	4816.37	5412.00	1106.70	3646.00	2444.00
Zhejiang	7020.99	3729.79	4455.45	1412.16	2826.74	1803.00

Source: China Statistical Press (2011) [5]

Table 3.3 Yields of major agricultural products in 2010 (Unit: kg/ha)

The average yield of rice in Shanghai, Shandong and The average yield of rice in Shanghai, Shandong and Jiangsu is more than 8000 kg per hectare, significantly higher than in other regions and the national average. Only Beijing's is lower than the national average of 6,553 kg per hectare. There are very small areas of rice cultivation in the three northern regions, and hence it is unnecessary to compare yield of rice in the north and the south. Wheat yields are above 5000 kg per hectare in Shandong and Hebei, while they are less than 4000 kg per hectare in Zhejiang and Shanghai, being lower than the national average. Overall the wheat yield in the three northern regions is lower than that of the three southern regions. The maize yields in Shanghai and Shandong are higher than the other regions and the national average, and the difference between the south and the north is not significant. Meanwhile, yields of cotton in Shanghai and Zhejiang are higher than in the other regions and the national average, and overall the three northern regions are higher than the three regions in the south. The average peanut yield in Shandong is much higher than that in other regions and the national average. The yield of rapeseed in Shandong is the highest, but the yields in Hebei and Beijing are very low, so overall the yield in the three northern regions is higher than the three regions in the south. The yield differences are likely to be influenced by regional climate, soil and irrigation.

3.4 Agricultural production, rural environment and public services

3.4.1 Demographic characteristics

The population sizes of villages are quite different, with the smallest value of 655 in Beijing, while the largest value of 3804 is sampled from Jiangsu Province. The villages in Hebei have the highest number of non-agricultural enterprises, followed by those in Shanghai Province. However, in terms of revenue from non-agricultural enterprises, the villages in Shanghai derive the largest value, followed by Zhejiang. To the nearest location of the local government from the village, the shortest is Shanghai, only 1 km; the farthest is Shandong, 29 km. The distance to the nearest market town from all villages is less than 7 km (Table 3.4).

Area (number of villages)	Population (person)	Number of households	Number of non-agro enterprises	Non-farm income (10000 yuan)	Annual cash income per farmer	Distance to nearest countryseat (km)	Distance to nearest town (km)
Beijing (3)	655	263	-	-	7019	15	5
Hebei (3)	2896	774	23	458	3583	21	3
Shandong (5)	779	213	3	54	4100	29	3
Shanghai (3)	3093	1006	18	6100	11092	1	1
Jiangsu (3)	3804	957	3	733	9400	12	6
Zhejiang (4)	1386	451	3	2230	9794	22	4

Note: - means no data; $100 exchange 644 yuan on July 29, 2011, http://www.boc.cn/sourcedb/whpj/

Source: field survey by the authors

Table 3.4 Characteristics of each sampled village on average

The farmers in 2010 with the greatest cash income reside in Shanghai, where the average per capita is 11092 yuan, followed by Zhejiang, Jiangsu, Beijing, Shandong and Hebei (Table 3.5). Farmers' cash income in our survey data is much lower than the data from the China Statistical Yearbook 2010. In Beijing in particular, it is only about half of the Yearbook's data. Nevertheless, except for Beijing, we can get the same order of the other regions on data of annual cash income from the two sources. Therefore, the survey can be deemed as being representative of farmers in present day China. In Table 3.5, we have listed the data of cash income per household from the farmers' survey, where farmers from Jiangsu Province are recorded as having most cash income.

Data source	Year	Beijing	Hebei	Shandong	Shanghai	Jiangsu	Zhejiang	Total
China Statistic Yearbook	2009	14198	6266	7445	15189	9738	12177	
Village survey	2010	7019	3583	4100	11092	9400	9794	
Farmer survey	2010	2.00*	2.14	2.40	2.99	3.27	3.07	2.65
Counts of farmers	2010	88	120	60	88	89	110	555

*Note: 1=less than 10 thousand yuan, 2=10-30 thousand yuan, 3=30-50 thousand yuan, 4=more than 50 thousand yuan

Table 3.5 Annual cash income of the farmers (yuan per capita)

3.4.2 Agricultural production

According to Table 3.6, the paddy field area and the orchard area in the south are greater than those in the north. The dry land area in the south is smaller than it in the north and five southern villages possess fish ponds, while there are no fish ponds in the northern villages. In addition, the average area of greenhouse in the south is found to be larger than that in the north.

	Paddy field		Dry land		Orchard		Fish ponds		Greenhouse	
	North	South	North	South	North	South	North	South	North	South
Mean	788	2321	1276	457	181	385	0	258	61	90
Max	1200	8000	4000	2000	300	1000	0	1200	300	220
Min	550	50	26	10	30	30	0	4	6	6
Std. D.	295	2450	1290	864	138	535	0	527	117	95
Sample	4	9	9	5	3	3	0	5	6	4

Note: north means Beijing, Hebei and Shandong; south means Shanghai, Jiangsu and Zhejiang; 1 mu equals to 1/15 hectare

Source: field survey by the authors

Table 3.6 Areas of agricultural land in each village (Unit: mu)

Rice is widely planted in the south, but none is planted in the sampled northern villages, despite there being paddy fields. The acreage of wheat in the south and north is almost the same. The acreage of corn in the south is much less than in the north. The vegetable acreage in the south is more than three times that of the north (Table 3.7).

	Rice		Wheat		Corn		Vegetable	
	North	South	North	South	North	South	North	South
Mean	0	1824	1146	1147	1173	223	113	388
Max	0	6000	2500	4000	3100	1000	300	680
Min	0	346	260	50	260	5	10	83
Std.D.	0	1753	823	1642	873	384	162	299
Sample	0	9	8	7	10	6	3	3

Source: field survey by the authors

Table 3.7 Average crop acreage in each village (Unit: mu)

The villages in the north did not plant rice in 2010. The yield of wheat in the north is a little more than in the south. The yield of corn in the north is much more than in the south. The yield difference of the survey data is the same as in the statistical data, according to Table 3.4. We cannot compare the yield of vegetables in the north and south because no data are available recording the vegetable varieties (Table 3.8).

	Rice		Wheat		Corn		Vegetable	
	North	South	North	South	North	South	North	South
Mean	-	507	382	322	444	298	2500	30650
Max	-	1000	500	500	1000	400	2500	90000
Min	-	256	200	150	175	90	2500	200
Std.D.	-	227	112	151	238	141	-	51404
Sample	-	8	8	6	10	6	1	3

Source: field survey by the authors

Table 3.8 Average yields of the major crops (Unit: kg/mu)

In the north only three villages have dairy cow farmers, while no villages sampled in the south answered as raising cows. As for swine, the number in the north is a little more than in the south. Meanwhile, the number of broiler in the north is almost twice that of the south (Table 3.9).

	Dairy cow		Swine		Broiler	
	North	South	North	South	North	South
Mean	36	0	1343	1247	6257	3386
Max	70	0	4000	5000	16500	10000
Min	4	0	140	180	200	200
Std.D.	33	0	1300	1876	6849	3534
Sample	3	0	7	6	7	7

Source: field survey by the authors

Table 3.9 Numbers of major livestock in the north and south

Similarly, although the maximum scales of dairy cow farmers are small in the north, no cow is raised in the southern villages interviewed. Scales of swine farmers (or companies) in the north are much larger than in the south. Scales of broiler farmers (or companies) in north are smaller than in the south (Table 3.10).

	Dairy cow		Swine		Broiler	
	North	South	North	South	North	South
Mean	22	0	372	194	2804	3645
Max	35	0	1000	400	6000	20000
Min	2	0	110	20	20	10
Std.D.	18	0	334	116	2274	7290
Sample	3	0	6	7	5	7

Source: field survey by the authors

Table 3.10 Max. number of livestock raised in one company or farmer

There are six villages where pollutant-free agricultural products are planted, eight villages where green agricultural products are planted, only one for organic agricultural products, and 11 villages where nothing of certification of agricultural products applies. There are no significant differences in certification of agricultural products between the south and north (Table 3.11). Meanwhile, the main methods and sources of agricultural irrigation are surveyed as quite different between the south and north. The main sources of agricultural irrigation in the south are rivers, lakes and ponds. There is a lot of water in the south, so the main methods of agricultural irrigation are soil channels and concrete channels. The main sources of agricultural irrigation in the north are wells, rivers and rain. There is not water enough in the south, so the main methods of agricultural irrigation are pipes and soil channels. The irrigation of soil channel is not efficient. The reason why the soil channel is one of the main methods of agricultural irrigation in the north may be that there is not enough money to build concrete channels (Table 3.12).

	Free pollutant agro-products		**Green agro-products**		**Organic agro-products**		**Nothing**	
	North	**South**	**North**	**South**	**North**	**South**	**North**	**South**
Number of villages	2	4	4	4	1	0	6	5
Total acreage (mu)	80	700 (3)	920	330 (3)	10	0	0	0

Note: two villages did not answer; numbers in parentheses indicate number of villages

Source: field survey by the authors

Table 3.11 Certification of agricultural products

	Soil channel		**Concrete channel**		**Pipe**		**Others**	
	North	South	North	South	North	South	North	South
Village	5	7	1	3	6	0	1	0
	Well		River		Lake and pond		Rain	
	North	South	North	South	North	South	North	South
Village	9	0	6	9	0	2	3	0

Source: field survey by the authors

Table 3.12 The main methods and sources of agriculture irrigation

3.4.3 Rural environmental situation

In the 21 sampled villages, there are 7 villages surveyed as having rubbish collection points (RCPs) in all the focal living points, accounting for 33% in total. Meanwhile, 8 villages have RCPs in most of their focal living points, with the proportion of 38% in total. In addition, there are 5 villages (24% in total) surveyed as having no RCPs all of which locate in Hebei and Shandong Province (Figure 3.5). Thus in general, situation in the south is better than the north. In 68% of the villages, it is the village committee who manages the rubbish collection. The two villages where nobody manages to collect rubbish are located in Shandong Province. Two villages in Hebei Province did not answer (Figure 3.6).

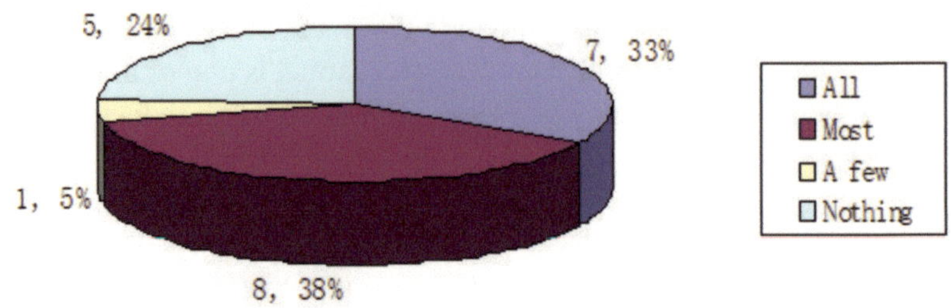

Figure 3.5 Situation of rubbish collection points

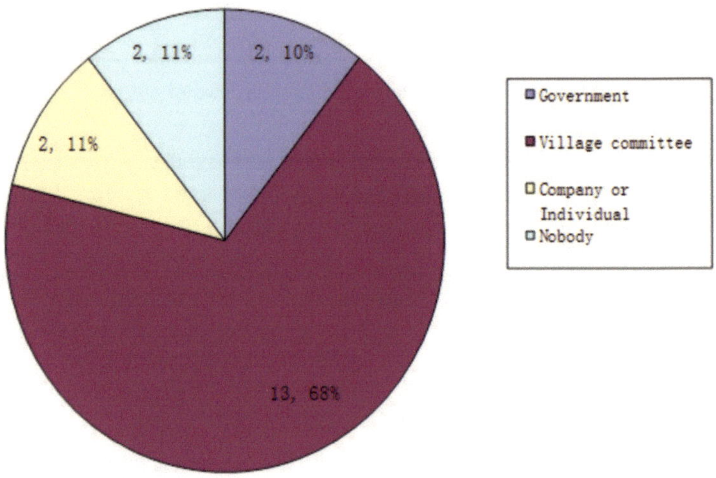

Figure 3.6 Managers of rubbish collection

The percentage of the collective rubbish disposal is almost consistent with the ranking of RCP. The five villages where the proportion of the collective rubbish disposal is lower than 10% are in Hebei and Shandong Province (Table 3.13).

		Whether there are RCPs?		
All	**Most**	**A few**	**Nothing**	
Village	7	8	1	5
		% of the collective rubbish disposal		
Percentage	100%	90-100%	70-90%	<10%
Village	8	4	3	5

Source: field survey by the authors

Table 3.13 The situation of rubbish collection

The proportion of the collective disposal of domestic sewage is 100% in six villages located in Beijing, Shanghai, Jiangsu and Zhejiang, while the ratio is less than 10% in the six villages of Hebei, Shandong and Zhejiang. In the 20 surveyed villages, there are 13 villages where more than 50% of domestic sewage is collectively disposed of. This illustrates that domestic sewage is handled well in survey areas. Of 19 surveyed villages, there are six villages where more than 70% of company wastewater is collectively disposed of, but less than 10% in 10 villages. In 20 surveyed villages, there are six villages where more than 70% of animal manure is collectively disposed of, but less than 10% in 10 villages. The villages where less than 10% of wastewater and manure are collectively disposed of are located across the six surveyed areas, i.e., not concentrated in one area. The pollution from company wastewater and animal manure is serious in the survey (Table 3.14).

			Domestic sewage				
Percentage	100%	90-100%	70-90%	50-70%	30-50%	10-30%	<10%
Village	6	2	3	2	0	1	6
			Company wastewater				
Percentage	100%	90-100%	70-90%	50-70%	30-50%	10-30%	<10%
Village	0	2	4	0	2	1	10
			Animal manure				
Percentage	100%	90-100%	70-90%	50-70%	30-50%	10-30%	<10%
Village	3	1	2	0	0	4	10

Source: field survey by the authors

Table 3.14 Percentages of collective wastewater disposal

The percentage of the farmers who had built methane tanks is less than 10% in 17 villages. As of 2009 the rural population stands at 34300 in Jiangsu Province. Given 3.5 persons per family there would be 9800 rural households. According to data there are 547 rural households who built methane tanks up to June 2009. So the percentage of village farmers who had built methane tanks would be 5.6% (Figure 3.7).

3.4.4 Rural public services

There are 10 villages where the agricultural service stations guide fertilizer use methods, of which seven are in the south. There are seven villages where the agricultural service stations guide pesticide use methods, of which five are in the south. Clearly, the agricultural service stations are dominant in fertilizer and pesticide use guidance, while the agricultural service stations are playing a greater role in the south than in the north. The government plays an important role in guiding livestock quarantining, and rubbish and domestic sewage disposal. Some villages choose others to supply the guidance on production and farmer's living, either because these villages have no corresponding agricultural production and living activities, or because there was nobody to provide these services on agricultural production and living (Table 3.15).

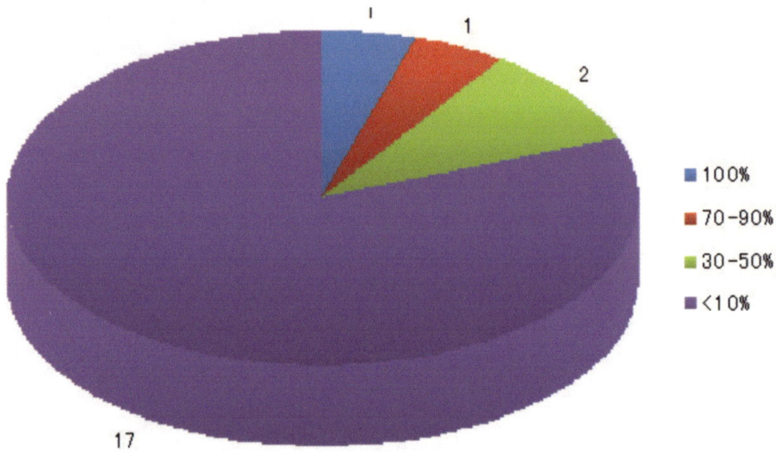

Figure 3.7 Ratio of villages with methane tanks

	Use of the fertilizer	Use of the pesticides	Livestock quarantine	Use of methane tank	Rubbish classification	Sewage discharge
Agricultural service station	10	7	1	3	1	1
Government	2	2	5	1	7	5
Sales company	1	1	0	0	0	0
Farmer technician	3	3	3	3	0	0
Others	4	4	5	6	4	6

Note: a section of the villages did not answer

Source: field survey by the authors

Table 3.15 Guidance on production and farmers' living in 2008-2010

As to the question of how many times guidance on agricultural production and farmers' living in 2008-2010 were carried out, there are only a few answers. From these limited answers, it appears that the guidance on agricultural production and farmers' living in 2008-2010 was carried out at least once a year, including the use of methane tanks (Table 3.16).

	Use of the fertilizer	Use of the pesticides	Livestock quarantine	Use of methane tank	Rubbish classification	Sewage discharge
Mean	3.4	3.5	7.5	2	7	4.5
Max	6	6	12	3	8	6
Min	2	2	3	1	6	3
Village	5	4	2	2	2	2

Source: field survey by the authors

Table 3.16 Number of guidance on agricultural production and farmers' living in 2008-2010

In terms of getting fiscal subsidies among the 19 villages, eight villages received subsidies to promote the use of methane tank; nine villages received subsidies to advocate the reduced use of fertilizer, while four villages were subsidized for reduction of pesticides. This indicates that fiscal subsidies are helpful in improving the rural environment, from the perspectives of adopting methane tanks, reducing the use of fertilizers and pesticides. Meanwhile, Table 3.17 shows that only two villages received fiscal subsidies for using biodegradable plastic sheets; four villages did not receive any subsidies - located in Jiangsu, Zhejiang and Shandong Provinces. Now that many farmers use plastic sheets in agricultural production, further fiscal subsidies on biodegradable plastic sheet usage will play a more important role in improving the rural environment.

	Methane tank	Reduction of fertilizer	Reduction of pesticide	Biodegradable plastic sheet	Self-made silage	Nothing
Village	8	9	4	2	1	4

Source: field survey by the authors

Table 3.17 Objectives of the government subsidies

As shown in Table 3.18, transfer payment, collective economy income and agricultural subsidies are the main three sources of public expenditure in villages. In this survey, six villages get transfer payments from a higher level of government which is more than 60% of their total expenditure; seven villages have collective economy income which is more than 60% of their total expenditure; 14 villages have transfer payments and collective economy income which is more than 50% of their total expenditure. Meanwhile, the agricultural subsidies are revealed as accounting for only a smaller proportion of total expenditure. The transfer payment could improve the rural environment, but we are not sure that if the collective economy income can do the same. In addition, the village's companies are interviewed as providing funds to the village, although some of them may pollute the environment simultaneously.

	Transfer Payment	Agricultural subsidy	Collective income	Corporate sponsorship	Social sponsorship	Villagers funding	Others
Village	14	11	13	2	5	4	1

Source: field survey by the authors

Table 3.18 Source of public expenditure in villages

3.4.5 Relation of rural environment and economic level

In Figure 3.8, 1 means all, 2 means most, 3 means a few, and 4 means nothing. The correlation coefficient of GCP and cash income is -0.38. There are RCPs in all or most of the focal living points of villages when the annual cash income of per farmer is over 6000 yuan. The correlation coefficient of GCP and the distance from village to the nearest county is 0.49.

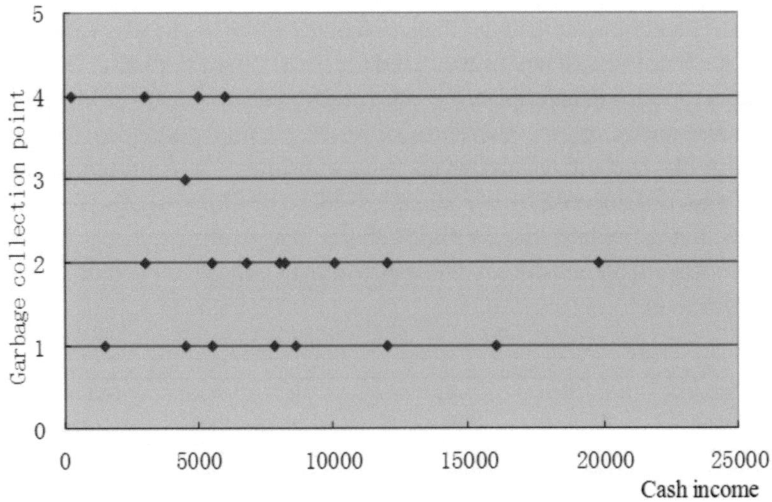

Figure 3.8 Relationship of GCP and cash income

In Figure 3.9, we use a similar system of 1means 100%, 2 means 90-100%, 3 means 70-90%, 4 means 50-70%, 5 means 30-50%, 6 means 10-30%, 7 means <10%. The correlation coefficient of cash income and the collective domestic sewage disposal is -0.30. The correlation coefficient of the collective domestic sewage disposal and the distance from village to the nearest county is 0.42.

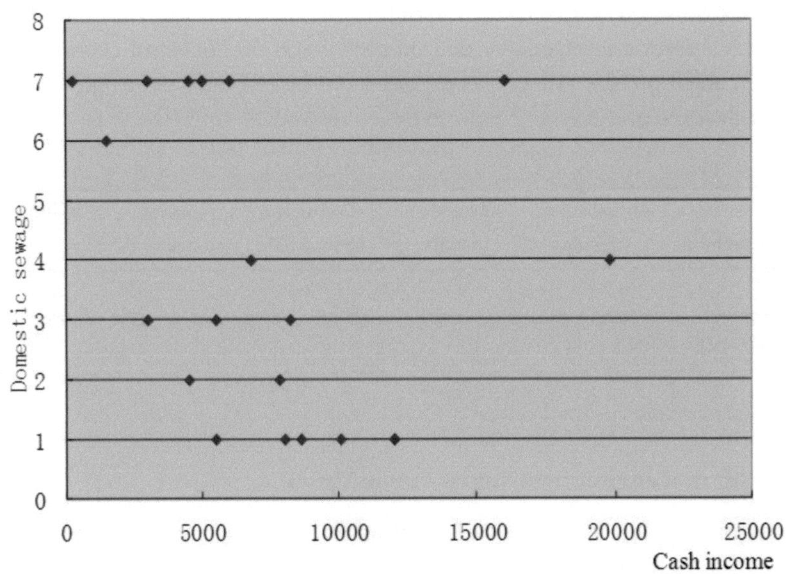

Figure 3.9 Relationship of domestic sewage and cash income

3.5 Conclusion

Land type, crop acreage, irrigation water source and irrigation methods have shown a certain difference between the north and the south. The proportion of the collective disposal of rubbish and sewage in rural areas is higher than industrial sewage and animal manure. The agricultural production and farmers' life guidance services are primarily supplied by government and agricultural extension centres. In about 50% of surveyed villages the government supplies subsidies for methane tanks and reduction of fertilizers, but only 10% in the case of biodegradable plastic sheets. In the 21 surveyed villages, 14 of them have transfer payments and collective economy income which is more than 50% of their total expenditure. The ratio of the collective rural rubbish and sewage disposal is positively correlated with income level of the farmers, and is negatively correlated with the distances from the towns.

References

[1] Wei, J, et al. A review of research on pollution prevention and control of rural environment in China, Ecology and Environmental Sciences (2010). (in Chinese).

[2] Wang, F. Study on the rural residents' environmental awareness, Journal of Qingyuan Polytechnic (2010). (in Chinese).

[3] Tang, L, & Zuo, T. Survey and analysis on the pollution condition of rural area in China: evidence from 141 villages, China Rural Survey (2008). (in Chinese).

[4] Huang, J, & Liu, Y. Environmental pollution in rural China and its driving forces, Chinese Journal of Management (2010). (in Chinese).

[5] CSP (China Statistical Press) China Statistical Yearbook (2011). http://www.stats.gov.cn/tjsj/ndsj/2011/indexch.htm.

Chapter 4

Farmer Behaviours Toward Food Safety and the Agro-Environment

Dongpo Li, Tinggui Chen, Hui Zhou and
Teruaki Nanseki

As revealed in the above chapters, increasing application of agricultural chemicals, including chemical fertilizers and pesticides, has been demonstrated as a key factor in improving China's agricultural productivity over the last few decades. Meanwhile, the excessive use of chemicals has resulted in serious threats and losses to the ecological environment, human health and economic development. As household farms are the overwhelming managerial units in Chinese agriculture, this chapter tries to accelerate the safe application of pesticides, through understanding the behaviours and determinants of the farmers.

4.1 Application of agricultural chemicals over the last few decades

In 2009, the chemical fertilizer applied to agriculture amounted to 54.04 million *tons* in China and maintained an average annual growth rate of 6.01% since 1978 [1]. Meanwhile, fertilizer consumption was 467.98 kg per hectare of arable land in 2008, much larger than the average amount of 134.93 kg per hectare amongst 175 countries [2]. A field test has revealed the low fertilization efficiency in China: the average nitrogen absorption efficiency of wheat, corn and rice are 28.3%, 28.2% and 26.1%, far lower than that of 40-60% in European and American countries [3]. Furthermore, even lower nitrogen absorption efficiency of only 10 % exists in vegetables, fruits and flowers [4]. According to the Bulletin of the First National Census on Pollution Sources issued in 2010, the non-point pollution (NPP) of agriculture has become the first source of water contamination in China, while chemical fertilizer applied in crop production constitutes the main source of agricultural NPP. The large volume of fertilizer residues has become a major source of environmental pollution and food safety incidents, thus proper application of fertilizer is drawing unprecedented public attention. The Chinese government has adopted the control of agricultural NPP into the twelfth Five-year Plan (2011-2015), with strengthening regulations on fertilizers.

By the end of 2010, the total amount of chemical pesticides produced in China amounted to 2.34 million tons, maintaining an average annual growth rate of 10.32% since 1985 [1]. China has become the largest producer, user and exporter of pesticides in the world [5]. Meanwhile, the improper use of pesticides has become a major source of food safety incidents, which have resulted in serious threats and losses to the ecological environment, human health and economic development. Therefore, safe application of pesticides is drawing unprecedented public attention and Chinese government is strengthening regulations on the production, marketing and use of pesticides [6].

4.2 Theoretical models

4.2.1 Model on fertilizer application

In this study, only those farms that answered as using fertilizer in 2010 are included, thus the sample consists of 294 valid responses from this survey. Based on the theoretical model specified above, we include 10 indicators to represent the demographic characteristics of each farm (Table 4.1). Simultaneously, these indicators will be used as candidate determinants to interpret farmers' behaviours on fertilizer application.

(1) Considering the importance of householders in making productive decisions within household farms, many studies included concerning variables as determinants in the analysis of safe agricultural production. In this study, we include three variables to describe attributes of the householders, i.e., human resources, as *gender* [7], *age* [8] and education level (*edu*) [9]. (2) In agrarian societies, land is not used solely as a means by which to generate a livelihood, it is often also used for accumulating wealth and transferring it between generations [10]. Thus two continuous variables on land cultivation are introduced: the sowing area of total agricultural products (*scale*), rather than total area of farmland is adopted with the consideration of multiple cropping [11]; sowing ratio of grain crops (*grainr*) is included to identify the effects of land use structure. (3) Meanwhile, another two variables are introduced to measure impacts of discrepancies in household income: total annual cash income (*income*) affects household budgets and thus inputs to agriculture, including the purchase of fertilizer [8-9]; ratio of income from migrant job (*mir*) shows the main sourcing structure of household income, which affects the relative importance of agriculture and also the corresponding inputs [12]. (4) To model the influence of geographic location on farmers' application of fertilizer [13], two dichotomous dummy variables are included with *north* (north or south of China) equal to 1 if a farm is from Beijing, Hebei or Shandong, and *metro* (metropolises or not) coded as 0 if a farm affiliates to neither Beijing nor Shanghai. The statistical summary of each variable is shown in Table 4.1. Finally, according to the China National Fertilization Regionalization[1] [14-15], the sampled areas cover four sub-regions as shown in the statistics following the characteristic variable of *fregion*.

1 The China National Fertilization Regionalization is drafted by the Soil and Fertilizer Institute, Chinese Academy of Agricultural Sciences. According to the soil condition and fertilization characteristics, this national planning divides China's farmland into 31 sub-divisions within eight divisions.

Characteristic	Unit	N	Mean	Min	Max	Std. D.	C.V.
Age of householder (age)	year	288	50.368	26.000	85.000	10.708	0.213
Sowing area (scale)	mu [a]	294	5.341	0.100	38.000	5.163	0.967
Ratio of grains sowing scale (grainr)	%	289	36.845	0.000	100.000	36.702	0.996
Ratio of migrant income (mir) [b]	%	281	35.362	0.000	100.000	41.297	1.168
Gender of farm head (gender)	dummy	288	1=male (275 c); 0=female (13)				
Education level of farm head (edu)	dummy	282	1=illiteracy (12); 2=primary (71); 3=middle (149); 4=high (42); 5=advanced (8)				
Total cash income in 2010 (income)	dummy	291	1=under 10000 yuan (42); 2=10000-30000 yuan (105); 3=30000-50000 yuan (88); 4=over 50000 yuan (56)				
North or south of China (north)	dummy	294	1=north (141); 0=south (153)				
Metropolises or not (metro)	dummy	294	1= Beijing or Shanghai (74); 0=other regions (220)				
Fertilization region (fregion)	dummy	294	1=Yanshan-Taihang mountainous areas (33); 2=Yellow-Huaihe-Haihe Plain (82); 3=Yangtze River plain (106); 4=Foothill areas South of Yangtze River (73)				

Note: [a] as a main unit of land measurement in China, 1 mu=666.67m^2; [c] the income sources contain migrant jobs and sales of agricultural products; [c] the bracketed numerals denote counts of farms.

Source: field survey by the authors

Table 4.1 Demographic characteristics of the sampled farms applying fertilizer

4.2.2 Model on pesticides application

Here, only farms that answered as using pesticides in 2010 are included in this chapter. From this survey, a total sample size of 220 valid responses is used in this study. We include nine indicators to represent the demographic characteristics of each farm (Table 4.9). In the following sections, these indicators will be used as candidate determinants to interpret farmers' behaviours.

(1) Due to the key role of the householder in making productive decisions within a family farm, many studies included relevant variables in the analysis of safe agricultural production. In this study, we include three variables to describe characteristics of the householders, i.e., human resources, such as *gender*, *age* and education level. (2) At the same time, to model the impacts of land cultivation to safe agricultural production as in Song et al. (2010) [16], two continuous variables are introduced: the sowing area of total agricultural products (*scale*), rather than total

area of farmland is adopted with the consideration of multiple cropping [14]; sowing ratio of grain crops (*grainr*) is included to identify the significance of land use structure. (3) Meanwhile, another two variables are used to measure the impacts of discrepancies in household income: total annual cash income (*income*) affects household budget and thus the inputs to agricultural production, including those on pesticides and spraying apparatuses [17-18]; ratio of income from migrant job (*mir*) shows the main sourcing structure of family income, which affects the relative importance of agriculture and the inputs [12]. (4) Finally, two dichotomous dummy variables are incorporated to show the importance of geographic location as in Zhang et al. (2004) [19], with *north* equal to 1 if a farm is from Beijing, Hebei or Shandong, and *metro* coded as 0 for farms locating in neither Beijing nor Shanghai.

4.3 Analysis on fertilizer application

4.3.1 Behaviours on fertilizer application

To capture the major behaviours of chemical fertilizer application in a farm, in addition to an aggregate amount, quantities of nitrogen, phosphate, potash and compound fertilizers used in each agro-product are included.

In the sampled farms, the nitrogen fertilizers mainly include carbamide, ammonium bicarbonate, etc.; the major phosphate fertilizer used is calcium superphosphate; potash fertilizers consist of potassium sulphate, etc. Amongst the three types of macro-element fertilizers, nitrogen fertilizers are most widely used by 278 (94.56%) farms, while potash fertilizers are used only in four (1.36%) farms. Although many compound fertilizers contain all the macro elements, the general fertilizing trend of rich nitrogenous and poor potash nutrients is testified to from the survey data (see [7, 14]). Meanwhile, the application of organic fertilizer (mainly including manure and compost) is represented in terms of the counts of farms amongst both the total sample and those who used chemical fertilizer simultaneously (Table 4.2).

	Unit	N	Mean	Min	Max	Std. D.	C.V.
Chemical fertilizer	kg/mu	294	58.489	8.890	285.710	45.609	0.780
Nitrogen	kg/mu	278	34.589	2.140	285.710	34.276	0.991
Phosphate	kg/mu	12	29.748	12.820	85.710	20.664	0.695
Potash	kg/mu	4	132.500	10.000	200.000	89.954	0.679
Compound	kg/mu	194	36.714	4.440	200.000	28.065	0.764
Organic fertilizer used in total farms	Dummy	300		1=used (206); 0=unused (94)			
Farms that used organic and chemical fertilizers	Dummy	224		1=used (137); 0=unused (87)			

Note: the bracketed numerals denote counts of farms

Source: field survey by the authors

Table 4.2 Application of fertilizer in the sampled farms

	Application of chemical fertilizer						
	Unit	N	Mean	Min	Max	Std. D.	C.V.
Wheat	kg/mu	120	63.928	2.330	333.330	40.107	0.627
Corn	kg/mu	120	51.159	8.330	175.000	34.749	0.679
Rice	kg/mu	61	46.505	6.670	220.000	35.997	0.774
Cotton	kg/mu	32	76.189	5.000	266.670	64.482	0.846
Fruiter	kg/mu	9	104.153	50.000	285.710	77.292	0.742
Vegetable	kg/mu	51	120.415	10.000	400.000	101.775	0.845
Oilseed	kg/mu	41	82.603	10.670	190.000	35.687	0.432
Peanut	kg/mu	25	62.799	6.000	140.000	33.240	0.529

Source: field survey by the authors

Table 4.3 Application of fertilizer in each agricultural plant

The surveyed agro-products include wheat, corn, rice, cotton, fruits, vegetables, oilseed and peanut. The average fertilizer used in the three main grain crops of wheat, corn and rice is 55.31 kg per mu, which is much less than that of the other products at 91.60 kg per mu (Table 4.3). Within the three main grain crops, wheat is applied with the largest amounts of fertilizer, while vegetable is mostly fertilized amongst all the other categories of agricultural plants. As to the organic fertilizer, it is much more widely used in the three main grain crops than in the other agricultural plants.

4.3.2 Perceptions on fertilizer application

In the questionnaire, four questions concern farmers' perceptions on fertilizer application, from choosing, applying and determining the amounts of chemical fertilizer, to the consequences of over fertilization. Moreover, as most of the fertilizer bags are made from PVC, containing a variety of toxic cancer-causing substances, long-term storage of food can easily bring about damp mildew and produce a strong carcinogen of aflatoxin [20]. Thus the improper disposal of fertilizer containers may endanger environmental safety and human health, and farmers' disposal of the used fertilizer packages is required simultaneously. For each question, the number of valid responses, counts and proportions of responses to each choice are shown in Table 4.4.

For most of the farmers, productive effects are the first determining factors in choosing and using fertilizer, less attention is paid to the environmental effects and sprayers' health. When determine the mounts of fertilizer, more than 50% of farmers answered that they followed the

1. Determinants on choosing of fertilizer (Single-choice with 546 valid responses)

Price	Productivity	Sellers	Peer practices	Follow-up services	Environment
103 (18.86%)	380 (69.60%)	16 (2.93%)	31 (5.68%)	1 (0.18%)	15 (2.75%)

2. Determinants of using fertilizer (Single-choice with 546 valid responses)

Costs	Productivity	Environmental effect	Sprayers' health	Quality of agro-product
120 (21.98%)	343 (62.82%)	13 (2.38%)	7 (1.28%)	63 (11.54%)

3. Determinants of fertilizing amounts (Single-choice with 546 valid responses)

Container instructions	Private experience	Extension instruction	Peer practices
278 (50.92%)	191 (34.98%)	42 (7.69%)	35 (6.41%)

4. Disposal of the fertilizer packages (Single-choice with 555 valid responses)

Rinsing and recycling	Burning up	Littering	Collective recycling	Others
326 (58.74%)	33 (5.95%)	57 (10.27%)	133 (23.96%)	6 (1.08%)

5. Consequences from over fertilization (Multiple-choice with 557 valid responses)

Crop lodging	Soil compaction	Water contamination	Yields increasing	Unknown	Others
255 (45.78%)	384 (68.94%)	148 (26.57%)	85 (12.56%)	39 (7.00%)	17 (3.05%)

Note: numerals are the counts of valid responses and the bracketed numbers are the corresponding %s of responses

Source: field survey by the authors

Table 4.4 Perceptions concerning fertilizer application

package instructions, while one third of them relied on their own experiences. In terms to the disposal of used fertilizer packages, almost 60% of farmers answered that they rinsed and reused, thus posing threats to the environment and human health. In many rural areas, farmers are storing their grains and other food stuffs in the used fertilizer bags, hence putting their food at high risk of being contaminated. Some farmers even rinse the used fertilizer bags in rivers, lakes, etc., hence constituting public water contaminations [21]. On the possible consequences of over fertilization, as a multiple-choice question, soil compaction is chosen by farmers accounting for an overwhelming ratio of 68.94%, followed by another choice of crop lodging with 45.78%. As to water contamination, this is chosen by less than one third of the respondents.

Thus, the proper and traditional perceptions coexist amongst the farmers, such as applying fertilizer according to the package instructions, concern about possible soil compaction due to over fertilization, while rinsing and reusing the packages for food storage, etc.

4.3.3 Analysis of the behaviour determinants

4.3.3.1 Calculating the Fertilization coefficient

Application of fertilizers is mainly affected by three factors: soil properties represented by the geographical location in the National Fertilization Regionalization, agricultural planting structure and farmers' propensities. This study aims to identify the discrepancies amongst farmers in terms of their propensities and thus behaviours with regard to fertilizer application. Hence, for further analysis, it is necessary to isolate the impacts of the former two factors. In this survey, average amounts of fertilizer applied per mu in each agro-product vary amongst different areas in the China National Fertilization Regionalization (Table 4.5).

To show the pure effect of farmers' propensities on determining amounts of chemical fertilizer, an indicator of *FC* (Fertilization Coefficient) for the *i*-th farm is formulated as:

$$FC_i = \sum_{j=1}^{8}\left(\frac{s_{ij}}{s_i}\cdot\frac{f_{ij}}{f_{\overline{kj}}}\right)\ (i=1,\ \cdots,\ 294; k=1,\ \cdots,\ 4) \tag{4-1}$$

Sub-division region	Wheat	Corn	Rice	Cotton	Fruiter	Vegetable	Oilseed	Peanut
Yanshan and Taihang mountainous areas		40.355			285.710			
Yellow river-Huaihe river-Haihe river Plain	64.975	55.229		104.729	118.890	157.097		51.528
Yangtze River Plain	60.485	53.274	54.526	58.5525		84.841	98.844	71.655
Foothill Areas South of Yangtze River			41.302	17.500	59.000	68.75	43.353	

Source: field survey by the authors

Table 4.5 Average amounts of fertilizer applied to each agro-product in different regions (Unit: kg/mu)

where s_{ij} is the sowing scale of the *j*-th agricultural product in the *i*-th farm; s_i is the total sowing scale of agricultural plants in the *i*-th farm; f_{ij} is the fertilizer applied per mu to the *j*-th agricultural product in the *i*-th farm; $f_{\overline{kj}}$ is the average amount of fertilizer applied per mu to the *j*-th agricultural product in the *k*-th region.

The summary statistics of the *FC*s for the 294 valid responses are shown in Table 4.6. To differentiate farmers' behaviours of fertilization driven by their propensities, they are divided into three groups in terms of their *FC*s, and the summary statistics for each group are provided in the same table. Group II embraces *FC*s fluctuating within 50% around 1, which represents the moderate amount of fertilizer determined by certain location and planting structure. Meanwhile, farms falling into Groups I and III indicate propensities of applying fertilizer with

50% under and over the moderate amounts, respectively. Statistics in this table show that Group II includes 180 farms (61.22%) with smallest Coefficient of Variance (CV) compared to the other two groups.

Group	Range	N	Mean	Min	Max	Std. D.	C.V.
I	(0, 0.50)	84	0.310	0.080	0.496	0.107	0.343
II	[0.50, 1.5)	180	0.949	0.500	1.486	0.252	0.266
III	[1.5, +∞)	30	2.133	1.514	3.804	0.588	0.276
Total		294	0.887	0.080	3.804	0.577	0.650

Source: field survey by the authors

Table 4.6 Summary statistics of FC in different groups

4.3.3.2 On the total amounts of fertilizer

To model the factors significant for the *FC* of a farm falling to any of the groups above, the dependent variable is a dichotomous indicator coded 1 if belonging to a certain group and 0 if not. As the OLS models are inappropriate for the discrete and limited dependent variables [22], a binary logit regression model is adopted and formulated as [23]:

$$Log\left[\frac{P(Y_1)}{P(Y_0)}\right] = \beta_0 + \sum_{i=1}^{9} \beta_i x_i + \varepsilon \tag{4-2}$$

where $P(Y_1)$ denotes the odds of *FC* belonging to a certain group, while $P(Y_0)$ represents being in other groups; x_1, x_2, ..., x_9 are the variables except for *fregion* in Table 4.1; β_0 and β_i are coefficients to be estimated; ε is the random error.

Estimation of the model is carried out through application of the binary logistic regression procedure in SPSS 13.0. The backward approach is adopted to remove the statistically insignificant variables, (p-value≥0.1), from the initial model with all the candidate determinants as independent variables. The final model selected includes predictors embracing a p-value less than 0.01 (Table 4.7). The column of B estimates log-odds coefficients of β_i in Eq. 4-2, for predicting the dependent variable from the independent variables. The last column lists the exponentiation of B, the ratio of $P(Y_1)$ and $P(Y_0)$, thus called *odds ratios* simultaneously. In this case, an *odds ratio* over 1 denotes that the farm is more probable to fall into the group, while an *odds ratio* less than 1 implies that the farm is more likely to fall out of the group [24].

The results show that (1) sowing area (*scale*) is an essential factor occurring in all the three groups, as negative within both I and III, while positive in Group II. This reveals the existence of scale economy in terms of fertilizer application in the sampled farms, thus the increase of managerial scale is favourable for appropriate fertilization [15]. (2) As another significant determinant, total annual *income* is beneficial for the probability of using fewer amounts of fertilizer [13]. In this survey, an apparent positive relationship exists between annual *income* and non-agricultural ratios. Within the farms included in this model, no migrant income occurred in the farms with annual cash income less than 10000 yuan, while this ratio in the other three income levels of Table 4.1 are 26.71%, 51.36% and 55.54%, respectively. The more non-agricultural income usually results in less farming time and attention to agricultural yields, thus the application of fertilizer may be decreased. (3) Meanwhile, effect from income ratio of migrant job (*mir*) is measured as being negatively correlated with amounts of fertilizer. Due to the instability and high expenditure of living away from homeland, most migrant farmers have to leave their families at home and engage in agriculture. As most of the left family members are women, children and the elderly, they are prone to improving agricultural productivity through the use of chemical fertilizer. The negative effect of *mir* in Group I may reveal that the more they get from migrant jobs, the more they can afford to use fertilizer [7]. Meanwhile, due to the lack of prime labour, most of them are not over-fertilizing, thus being positive in Group II. (4) The positive effect of *age* in Group I reveals that farms headed by the elderly are less prone to using fertilizer than the average amounts. This may be interpreted as due to limitation of physical power, disposable income, etc. (5) As analysed above, the three types of staple grain crops are supplied with less fertilizer than the other agricultural products. Therefore, their sowing ratios (*grainr*) go positively in Group I and are hence negatively correlated with the total amounts of fertilizer.

Group	Variable	B	S.E.	Wald	df	Sig.	*odds ratio*
I	Age of farm head (*age*)	0.020*	0.012	2.931	1	0.087	1.020
	Total cash income in 2010 (*income*)	0.560***	0.161	12.050	1	0.001	1.751
	Sowing area (*scale*)	-0.087***	0.032	7.275	1	0.007	0.917
	Sowing ratio of grain crops (*grainr*)	0.773**	0.384	4.054	1	0.044	2.166
	Ratio of migrant income (*mir*)	-0.009**	0.004	6.109	1	0.013	0.991
II	Total cash income in 2010 (*income*)	-0.444***	0.144	9.423	1	0.002	0.642
	Sowing area (*scale*)	0.098***	0.030	10.788	1	0.001	1.102
	Ratio of migrant income (*mir*)	0.010***	0.003	7.672	1	0.006	1.010
III	Sowing area (*scale*)	-0.115**	0.060	3.741	1	0.053	0.891

Omnibus tests of coefficients for model I: Chi-square (5)=23.941, Sig.=0.000***

Omnibus tests of coefficients for model II: Chi-square (1)=5.002, Sig.=0.025**

Omnibus tests of coefficients for model III: Chi-square (3)=25.191, Sig.=0.000***

Note: *** and **represents statistical significance in the level of 1% and 5%, respectively

Software: SPSS 13.0

Table 4.7 Binary logistic regression on FC of different groups

4.3.3.3 On the application of organic fertilizer

With the same binary logistic regression procedure in SPSS 13.0, we measure significant factors for the application of organic fertilizer in the sampled farms. Besides the aforementioned nine variables, we add three variables into the candidate determinants: amount of chemical fertilizer (*fert*), quantity of *livestock* and *poultry* to capture the possible impacts from these predictors, with the hypothesis that these variables affect farmers' application of organic fertilizer.

As shown in Table 4.8, through the predictor selection backward method, six variables are included in the final model. Judging from the *odds ratio* of each variable, impact of each variable can be identified.

(1) Farms from the north (*north*=1) are less likely to use organic fertilizer. Further investigations are necessary to explore the possible reasons for this from planting structure, habits and awareness on the function of organic fertilizer. (2) *Age* of farm head is positively correlated with farmyard application. This may be interpreted as the accumulation of social experiences, farmers are more confident about the effectiveness of organic fertilizer, or the significance of properly disposing of faeces and urine. (3) Similar with the findings of Zheng (2010) [25], annual cash *incomes* is also positively correlated with farmyard application. With the increase of income, farmers need cleaner environments and safer food supplies, thus they are apt to fertilize their farmland with organic fertilizer, rather than chemical fertilizer (as analysed above). (4) Farms with larger sowing *scale*s are less prone to use organic fertilizer, probably due to the fact that they are pursuing higher production efficiency and tend to use chemical fertilizer. In addition, the collection and application of organic fertilizer enough for their large sowing scales is consuming labour and funds. (5) The sowing ratio of grain crops (*grainr*) is negatively correlated with the application of organic fertilizer, which can be interpreted as that most of the grains are sold while the economic agro-products will be consumed by the farmers themselves. Hence, they are tending to fertilize the economic crops with organic fertilizer which is labour-consuming, but deemed as salubrious by the farmers [26]. (6) Income the ratio of migrant job (*mir*) is found to be negatively correlated with the application of organic fertilizer. The main reason behind this may be the fact that farms which are more reliant on the non-agricultural income usually have less time for farming, much less fertilizing their farmland through organic fertilizer. Meanwhile, no significant relationships are detected between the application of organic fertilizer and chemical fertilizer (similar to Zheng (2010) [25]), breeding of livestock and poultry. This indicates the existence of a certain blindness about the application of organic fertilizer, which may bring about improper disposal of manure and compost, thus environmental pollutions.

4.3.4 Conclusions and Recommendations

4.3.4.1 Major conclusions

Based on a survey of 560 household farms in six eastern provincial level regions of China, this study explores farmers' behaviours, perceptions and determinants of fertilizer application. The behaviours involve total amount of chemical fertilizer and the use of organic fertilizer;

Variable	B	S.E.	Wald	*df*	Sig.	*odds ratio*
North or south of China (*north*)	-1.265***	0.484	6.820	1	0.009	0.282
Age of farm head (*age*)	0.029*	0.016	3.192	1	0.074	1.029
Total cash income in 2010 (*income*)	0.574***	0.217	6.974	1	0.008	1.775
Sowing area (*scale*)	-0.123***	0.037	11.301	1	0.001	0.884
Sowing ratio of grain crops (*grainr*)	-0.010**	0.005	4.218	1	0.040	0.991
Income ratio of migrant job (*mir*)	-0.026***	0.005	25.254	1	0.000	0.975
Cases included in analysis: 267; Missing cases: 33; Total cases selected: 300						
Dependent variable: whether organic fertilizer is used, with 178 cases = 1, and 89 cases = 0						
Omnibus tests of model coefficients: Chi-square (6)=86.382, Sig.=0.000***						

Note: ***, **and *represent statistical significance in the level of 1%, 5% and 10%, respectively.

Software: SPSS 13.0

Table 4.8 Binary logistic regression on application of organic fertilizer

farmers' perceptions range from choosing and field application, the consequences of over fertilization and disposal of the used packages. Logistic regression models are used to identify the significant determinants of their behaviours.

The survey shows that most farms are using nitrogen fertilizers, while potash fertilizers are used in a few farms. Compared with the other plants, less chemical fertilizers are used in the main grain crops of wheat, corn and rice. Judging from the Fertilization Coefficient, more than 60% of farms are using fertilizer in amounts deviating by no more than 50% from the average amounts in each fertilizing region. Perceptions of proper fertilization are held by some farmers, including applying the fertilizer by instructions, recycling the packages collectively and concern about the possible crop lodging and soil compaction due to over fertilization. Simultaneously, traditional conceptions are still held by many farmers, such as the over emphasized production effects and private experiences, reusing the packages for food storage, etc.

According to the empirical analyses, sowing area and ratio of migrant income is measured as positively correlated, while annual income is negatively correlated with appropriate fertilization. As to the odds of using organic fertilizer, no significant effect is detected from chemical fertilizer application and the breeding of livestock and poultry. Nevertheless, cash income and age of householders are positively correlated with the application of organic fertilizer, while location in the north, sowing scale, ratio of grain crops and migrant income are measured as negatively correlated with the same behaviour.

4.3.4.2 Policy recommendations

(1) As shown above, the fertilizing elements are not well balanced, and amounts of fertilizer used in many farms deviate much from the moderate levels. Therefore, it is an urgent task for the government to provide prompt, accurate and convenient soil testing techniques, and recommend referential standardized fertilizing amounts to farmers with different land properties and planting structures [3]. (2) Enlarging the managerial scales of agriculture. As analysed above, larger scale is positively correlated with maintaining appropriate fertilizing amounts. Managerial scales of the farms can be expanded either through the concentration of land based on the farms' own wishes or joining into the Specialized Farmers' Cooperatives as demonstrated by Sun (2008) [27] and Dai (2010) [12]. (3) Promoting migrant employment of rural labour, as ratio of migrant income is positively correlated with appropriate use of fertilization and application of organic fertilizer. As for accelerating the transfer of surplus labour from agriculture to the other sectors, thus increasing rural household incomes, the main tasks include promoting vocational training, perfecting the employment information networks and protecting the legal rights of the migrant workers. (4) Strengthening social education on scientific fertilization. This survey reveals that behaviours, including fertilizing by private practices, misusing the used packages, etc., still exist amongst many farmers, and their perceptions on safe application of fertilizers need to be improved. Hence education on appropriate amounts of fertilizer, balancing the elements, proper recycling of the used packages, etc., needs to be strengthened [26].

4.4 Analysis on pesticide application

4.4.1 Behaviours on pesticide application

To capture the major behaviours of pesticide application in a farm, three aspects are included in our questionnaire. As usually a variety of pesticides, with different pest control and environmental effects, are used in a farm, weights of pesticides applied in each agricultural product are summed to constitute the total amounts. Meanwhile, as the control of toxic pesticides and promotion of biological pest controls are of great importance for safe agricultural production, relevant characteristics are also included. The toxic pesticides incorporate methamidophos, furadan (carbofuran) and folimat [28]. According to the No. 199 Bulletin of the China Agricultural Ministry (2002), it is prohibited to apply methamidophos in agriculture and Furadan cannot be used on vegetables, fruit, tea and medicinal herbs. As another major toxic pesticide, Folimat has been banned in some regions including Zhejiang [29], Jiangsu [30], etc.

The bio-control methods of pests in agriculture are measures to eliminate insects, mites, weeds and plant diseases, etc., that rely on certain biological mechanisms of secretion, smell, predation, parasitism, herbivory, etc., thus reducing the use of chemical pesticides. For example, using the smell of onions to kill microbes causing black spike of wheat, intercropping beans in corn fields to attract beneficial insects that prey upon pests, raising ducks and fish in rice fields to control weeds, etc [31]. The application of pesticides and bio-control measures in the sampled farms are shown in Table 4.10.

Characteristic	Unit	N	Mean	Min	Max	Std. D.	C.V.
Age of farm head (age)	year	211	49.68	26.00	78.00	10.08	0.20
Sowing area (scale)	mu [a]	220	5.95	0.50	38.00	4.82	0.81
Ratio of grains sowing scale (grainr)	%	217	39.20	0.00	100.00	35.40	0.90
Ratio of migrant income (mir) [b]	%	216	37.76	0.00	100.00	41.59	1.10
Gender of farm head (gender)	dummy	211	1=male (204 c); 0=female (7)				
Education level of farm head (edu)	dummy	208	1=illiteracy (8); 2=primary (49); 3=middle (106); 4=high (40); 5=advanced (5)				
Total cash income in 2010 (income)	dummy	218	1=under 10000 yuan (16); 2=10000-30000 yuan (86); 3=30000-50000 yuan (82); 4=over 50000 yuan (34)				
North or south of China (north)	dummy	220	1=north (129); 0 =south (91)				
Metropolises or not (metro)	dummy	220	1= Beijing or Shanghai (55); 0= the other regions (165)				

Note: [a] as a main unit of land measurement in China, 1 mu=666.67m^2; [b] the income sources contain migrant jobs and sales of agricultural products; [c] the bracketed numerals denote counts of farms

Source: field survey by the authors

Table 4.9 Demographic characteristics of the sampled farms applying pesticides

	Unit	N	Mean	Min	Max	Std. D.	C.V.
Total amount	kg/mu	220	1.05	0.01	11.67	1.92	1.82
Toxic pesticides	kg/mu	105	0.51	0.01	7.27	1.25	2.43
Methamidophos	kg/mu	47	0.37	0.01	3.33	0.59	1.59
Furadan	kg/mu	14	3.74	0.33	13.33	3.88	1.04
Folimat	kg/mu	62	0.53	0.00	5.00	1.05	1.98
Bio-control of total farms	dummy	306	1= implemented (46); 0= unimplemented (260)				
Bio-control of farms used pesticides	dummy	168	1= implemented (32); 0= unimplemented (136)				

Note: the bracketed numerals denote counts of farms

Source: field survey by the authors

Table 4.10 Application of pesticides in the sampled farms

	Application of chemical pesticides							Number of farms using	
	Unit	N	Mean	Min	Max	Std. D.	C.V.	Toxic pesticides	Bio-control
Wheat	kg/mu	95	0.37	0.01	3.00	0.56	1.53	48	9
Corn	kg/mu	48	0.23	0.02	1.25	0.27	1.21	32	6
Rice	kg/mu	46	1.09	0.02	5.00	1.31	1.20	15	2
Cotton	kg/mu	28	1.44	0.02	6.00	1.69	1.17	17	0
Fruiter	kg/mu	26	5.29	0.40	11.67	2.50	0.47	0	27
Oilseed	kg/mu	34	0.58	0.15	2.13	0.43	0.75	33	4
Soy	kg/mu	27	0.32	0.10	0.50	0.09	0.28	26	0

Source: field survey by the authors

Table 4.11 Application of pesticides in each agricultural product

The agricultural products we surveyed include wheat, corn, rice, cotton, oilseed, soy and fruits, and application of pesticides per mu of each product is presented in Table 4.11. The average pesticides used in the three main grain crops of wheat, corn and rice is 0.51 kg per mu, which is much less than that of the other products at 1.79 kg per mu. Meanwhile, judging from the Coefficient of Variance (CV), amounts of pesticides used in these main grain crops show larger discrepancies than that of the other products. According to the survey, toxic pesticides are used in all the products except for fruit, amongst which methamidophos is used in rice and soy, Folimat is used in wheat, cotton, cole and cotton, while Furadan is used in cotton. Finally, bio-control methods are used on far fewer farms, covering most of the products other than cotton and soy.

4.4.2 Perceptions on pesticide application

Within this questionnaire, five questions concern farmers' perceptions on pesticide application, from choosing and field application to the withdrawal periods, and the possible consequences of overdosing. Moreover, as pesticide containers may be toxic and improper disposal may threaten environmental safety and human health [32], another question is adopted in this topic. For each question, the number of valid responses, counts of responses and the corresponding percentages to each choice are shown in Table 4.13.

This shows that for most of the farmers, productive effects are the most determining factors in choosing and using pesticides, less attention is paid to the environmental effects and sprayers' health. When determining the doses, almost 50% of the farmers answer that they follow the container instructions, while one third of them rely on their own experiences. Although more than 80% of the farmers have heard of the withdrawal period of pesticides, less than 20% of the farmers answered as being well known. In the disposal of pesticide containers, almost 40% of the farmers answered as littering, thus threatening the environment and human health. On the possible consequences of overdosing, the negative effects on sprayers' health, food safety and environment are recognized by more than half of the respondents simultaneously. Hereby, the coexistence of the proper and traditional perceptions is shown amongst the farmers.

4.4.3 Analysis on the behaviour determinants

4.4.3.1 On the total amount of pesticides

In the prior studies, multivariate OLS regression models are used to identify the significant determinants of pesticide application, as in Wang (2004) [11], Li et al. (2007) [17], etc. In this study, the model used to find the important factors of total chemical pesticide amount is formulated as:

$$Y = \beta_0 + BX + u \tag{4-3}$$

where Y is the total amount of pesticides applied per mu, $X=(x_1, x_2, \ldots, x_9)^T$ is a vector containing the nine variables listed in Table 4.12, β_0 and $B=(\beta_1, \beta_2, \ldots, \beta_9)$ are coefficients that need to be estimated, while u is the random error.

Through the multivariate linear regression process, with the independent variable selection backward method in the statistical software of SPSS 13.0, four significant determinants are chosen in the final model. The relevant statistics of the model are shown in Table 4.12. The significant values of F and t (p-value < 0.1) indicate the good fitness of this model[2].

Variables	Unstandardized Coefficients		Standardized Coefficients	t	Sig.
	B	Std. Error	Beta		
(Constant)	0.107	0.493		0.217	0.829
Metropolis or not (*metro*)	2.355***	0.219	0.605	10.747	0.000
Gender of farm head (*gender*)	0.846*	0.481	0.099	1.759	0.080
Income ratio of migrant job (*mir*)	-0.004**	0.002	-0.109	-1.975	0.050
Ratio of grain sowing scale (*grainr*)	-0.010***	0.003	-0.222	-3.996	0.000
Valid N=199; F=33.13, Sig=0.000***; R^2=0.406					

Note: dependent variable: pesticides used per mu; ***, **and *represent statistical significance in the level of 1%, 5% and 10%, respectively.

Software: SPSS 13.0

Table 4.12 Statistics of the significant determinants on total pesticide used per mu

2 Although a not very high R2 value of 0.406 is given in the table, it should not be used to judge the fitness of a model. The fact that R2 never decreases when any variable is added to a regression makes it a poor tool for deciding whether one or several variables should be added to a model. Low R2s in regression equations are not uncommon, especially for cross-sectional analysis. Thus, using R2 as the main gauge of success for an econometric analysis can lead to inaccuracies [33].

The results show that farms affiliating to the two metropolises of Beijing and Shanghai (*metro*=1), or headed by males (*gender*=1) are positively correlated, while ratios of income from migrant jobs (*mir*) and grains sowing scales (*grainr*) are negatively correlated with the amount of chemical pesticides applied per mu. (1) The coefficient of *metro* can be explained by the comparison of average pesticides used per mu and other indicators of the farms. Within the 199 farms included in this model, farms affiliating to the metropolises applied 2.64 kg of pesticides per mu with the sowing area of 3.84 mu on average, while the corresponding indicators in non-metropolis farms are 0.38 kg per mu and 6.70 mu, respectively, thus the former may have to maintain high yields through more application of pesticides. Simultaneously, the higher annual cash incomes in farms affiliating to the metropolises[3] enable them to input more in pesticides. However, we should notice that this discrepancy may threaten the environmental and food safety of the metropolises.

1. Determinants on choosing of pesticides (Single-choice with 546 valid responses)					
Price	**Productivity**	**Sellers**	**Peer practices**	**Follow-up services**	**Environment**
103 (18.86%) [a]	380 (69.60%)	16 (2.93%)	31 (5.68%)	1 (0.18%)	15 (2.75%)
2. Determinants of using pesticides (Single-choice with 546 valid responses)					
Costs	**Productive effect**	**Environmental effect**	**Sprayers' health**	**Quality of agro-product**	
120 (21.98%)	343 (62.82%)	13 (2.38%)	7 (1.28%)	63 (11.54%)	
3. Determinants of pesticides dose (Single-choice with 546 valid responses)					
Container instructions	**Private experience**	**Extension Instruction**	**Peer practices**		
278 (50.92%)	191 (34.98%)	42 (7.69%)	35 (6.41%)		
4. Withdrawal period of pesticides (Single-choice with 557 valid responses)					
Knows very well	**Knows fairly well**	**Knows a little**	**Unknown**		
97 (17.41%)	248 (44.5%)	105 (18.85%)	107 (19.21%)		
5. Disposal of the pesticide containers (Single-choice with 550 valid responses)					
Individual recycling	**Burning up**	**Littering**	**Collective recycling**	**Others**	
79 (14.36%)	73 (13.27%)	212 (38.55%)	182 (33.09%)	4 (0.73%)	
6. Consequences from overdosing of pesticides (Multiple-choice with 557 valid responses)					
Sprayers' health	**Food security**	**Pollution**	**Higher effectivity**	**Unknown**	**Others**
337 (60.50%)	423 (75.94%)	316 (56.73%)	105 (18.85%)	16 (2.87%)	9 (1.62%)

[a] Note: numerals are the counts of valid farm, and the bracketed numbers are the corresponding %s of farms

Source: field survey by the authors

Table 4.13 Perceptions concerning pesticide application

3 Using the codes of 1, 2, 3, 4 to denote the ascending income levels of Table 1, within the 199 farms included in this model, the mean in farms affiliating to the metropolises is 3.02, while that in the other farms is 2.48.

(2) As to the finding that male headed farms are applying more pesticides, this indicates that males are more concerned about the productive effects of farming activities and more able to spray large volume of pesticides because of pure physical power, as investigated by Li et al. (2007) [32]. (3) The negative effect of income ratio of migrant job is in line with Li et al. (2007) [17]. The greater the amount of non-agricultural income, usually results in less farming time and attention to agricultural yields, thus the application of pesticides may be decreased. (4) As analysed above, the three types of staple grain crops are supplied with less pesticides than the other agricultural products. Therefore, their sowing ratio goes negatively correlated with the total amount of pesticides.

4.4.3.2 On toxic pesticide application

As to the model factors significant for application of toxic pesticides defined above, the dependent variable is a dichotomous indicator coded 1 if applied and 0 if not. As the OLS modes like Eq.4-3 are inappropriate for discrete and limited dependent variables [22], a binary logit regression model is adopted [34], [12] and defined as [23]:

$$Log\left[\frac{P(Y_1)}{P(Y_0)}\right]=\beta_0+\sum_{i=1}^{9}\beta_i x_i+\varepsilon \qquad (4\text{-}4)$$

where Y is the application of toxic pesticides with $P(Y_1)$ denoting the probability of being applied, while $P(Y_0)$ means that toxic pesticides are unapplied, x_i (i=1, 2, ..., 9) are the nine variables listed in Table 4.14, β_0 and β_i (i=1, 2, ..., 9) are coefficients that need to be estimated, and ε is the random error.

Estimation of this model is carried out through application of the binary logistic regression procedure in SPSS 13.0. The backward approach is adopted to remove the statistically insignificant variables, (p-value≥0.1), from the initial model with all the candidate determinants as independent variables. The final model includes four predictors, all of which embrace p-values less than 0.01 (Table 4.14). Column B estimates log-odds coefficients of β_i in Eq.4-4, for predicting the dependent variable by the independent variables. The last column lists the exponentiation of B, the ratio of $P(Y_1)$ and $P(Y_0)$, thus called *odds ratios* simultaneously. In this case, an *odds ratio* over 1 denotes that the toxic pesticides are more probably used, while an *odds ratio* less than 1 implies that the toxic pesticides are less likely to be used [24].

Within the four significant variables listed in Table 4.14, *mir* is positively correlated with the odds of toxic pesticides being applied, while the other three variables are negatively correlated with the application probability of toxic pesticides on a farm. (1) Being the capital and largest city in China, respectively, and especially thanks to the hosting of the Olympic Games and World Expo, Beijing and Shanghai have adopted stringent regulations to prevent the use of highly toxic pesticides [35-36]. Therefore, the less probability of applying toxic pesticides and negative effect of *metro* can be interpreted. (2) For a farmer, the greater the income from migrant jobs means less time and attention for farming in general. However, due to unstable conditions and high living expenditure outside of the native place (which is usually defined using the county level), most migrant farmers have to leave their families at home and engage in agriculture [19]. As most of the left family members are women, children and the elderly, they are prone to control pests through the more efficient toxic pesticides. The positive effect of *mir* may reveal that the greater the income obtained from migrant jobs, the more they can afford to buy and use the toxic pesticides. (3) However, when were observe farms' cash *income* with units of dozens of thousand yuan as shown in Table 4.14, farms with an upper level of income tend to use less toxic pesticides as their major income comes from non-agricultural sectors[4]. Through tradeoffs with the probable efficient pest control by toxic pesticides, most of them may prefer to conserve the environment and food security. (4) Finally, as the three types of staple grain crops need less pesticide in general, the application of toxic pesticides is negatively correlated with the *grainr* simultaneously.

Variables	B	S.E.	Wald	df	Sig.	*odds ratio*
Metropolis or not (*metro*)	-2.507^{***}	0.607	17.051	1	0.000	0.082
Income ratio of migrant job (*mir*)	0.018^{***}	0.005	11.975	1	0.001	1.081
Total cash income in 2010 (*income*)	-0.755^{***}	0.251	9.019	1	0.003	0.470
Ratio of grain sowing scale (*grainr*)	-0.027^{***}	0.006	18.828	1	0.000	0.974
(Constant)	2.515	0.640	15.458	1	0.000	12.363

Cases included in analysis: 199; missing cases: 21; total cases selected: 220

Dependent variable: whether toxic pesticides are used, with 93 cases = 1, and 106 cases = 0

Omnibus tests of model coefficients: *Chi*-square (4)=71.642, Sig.=0.000^{***}

Note: *** represents statistical significance in the level of 1%

Software: SPSS 13.0

Table 4.14 Binary logistic regression on whether toxic pesticides used

4.4.3.3 On the adoption of biological pest-controls

With the same binary logistic regression procedure in SPSS 13.0 and the nine variables as the candidate determinants, we measure the significant factors for the implementation of biological pest control in the sampled farms.

As shown in Table 4.15, using the backward predictor selection method, three variables are included in the final model. Judging from the *odds ratio* of each variable, (1) farms from the north (*north*=1) or (2) affiliating to the two metropolises (*metro*=1) are likely to adopt biological measures. Within the farms answered as conducting biological pest control, 36 are from the north and 31 from the two metropolises, accounting for 78.26% and 67.39% with the valid number of 46, respectively. As to the positive significance of *metro*, this may be because that

4 Within the 199 farms included in this model, no migrant income occurred in the farms with annual cash income less than 10000 yuan, while this ratio in the other three income levels of Table 1 are 22.70%, 54.58% and 60.56%, respectively.

as previously mentioned, being the capital and largest city in China respectively, Beijing and Shanghai are making full use of their solid industrial foundation and advantages in technology, trade, information - making greater efforts to promote the research and production of low toxicity and environmentally friendly pesticides (see [37, 31]). As to the difference between the north and south, further investigations are necessary to explore the possible reasons for variations in cropping structure, farming habits, the degree of pest damage, etc. [18], hence searching for suitable countermeasures to extend biological pest controls in different regions. Meanwhile, (3) income ratio of migrant job (*mir*) is found negatively correlated with the introduction of biological pest controls. this may be because those farms rely more on non-agricultural incomes, thus usually have less time to attend to farming, much less controlling pests through biological methods.

Variables	**B**	**S.E.**	**Wald**	**df**	**Sig.**	***odds ratio***
North or south of China (*north*)	0.980**	0.572	2.929	1	0.087	2.664
Metropolis or not (*metro*)	3.403***	0.571	35.490	1	0.000	30.056
Income ratio of migrant job (*mir*)	-0.011	0.007	2.574	1	0.109	0.989
(Constant)	-3.239	0.570	32.273	1	0.000	0.039

Cases included in analysis: 274; Missing cases: 286; Total cases selected: 560

Dependent variable: whether biological pest controls are implemented, with 27 cases = 1, and 247 cases = 0

Omnibus tests of model coefficients: *Chi*-square (3)=47.607, Sig.=0.000***

Note: ***, **and *represent statistical significance in the level of 1%, 5% and 10%, respectively

Software: SPSS 13.0

Table 4.15 Binary logistic regression on implementation of biological pest control

4.4.4 Conclusions and recommendations

4.4.4.1 Major conclusions

The survey shows that pesticides used on the three staple grain crops are less than those used on other products, but there is more discrepancy amongst the farms. The toxic pesticides are applied in most of the products and some 50% of the sampled farms, while bio-control methods are used in only about one sixth of the farms. Perceptions on proper application of pesticides exist amongst some of the farmers, including applying according to the instructions on the containers, awareness on the withdrawal periods, collective recycling of the containers, concern about sprayers' health and food security. Simultaneously, traditional conceptions still influence many of them, such as the over emphasized importance of productive effects and private experiences, littering by incorrect disposal of pesticide containers, etc.

According to the empirical analyses, farms in the two metropolises and those headed by males are positively correlated, while ratios of income from migrant jobs and grain sowing scales are negatively correlated with the amounts of pesticides applied. With respect to the application probability of toxic pesticides, the income ratio of migrant job is positively correlated, while the other three variables of *metro, income* and *grainr* embrace negative effects. Farms' location, whether north or affiliating to the metropolises, is measured as positively correlated, while ratio of migrant income is negatively correlated with the odds of adopting biological pest controls.

4.4.4.2 Policy recommendations

(1) Extending advanced techniques to improve pesticidal efficiency and guarantee safe application of pesticides. In addition to the alternative techniques and products of toxic pesticides, biological pest-controlling techniques, techniques on efficient pesticide spraying, monitoring the residues, decomposing rubbish including pesticide containers, etc., are urgently needed by the farmers. (2) Severe inspection on the production, circulation and use of highly toxic pesticides, including the improvement of the licensing, registration and classification systems of pesticide production, establishing tracing back systems and cracking down on the illegal production and trafficking of highly toxic pesticides. (3) According to the foregoing analysis, the ratio of migrant income is negatively correlated with the amount of pesticides; total income is negatively correlated with the use of toxic pesticides. Therefore, continuing transfer of surplus labour from agriculture to the other sectors is still necessary, which can simultaneously improve the total income of rural households. The main tasks include promoting vocational training, perfecting the employment information networks and protecting the legal rights of migrant workers. (4) This survey reveals that behaviours like littering from the incorrect disposal of containers and spraying pesticides by private practice still exist amongst many farmers, and their perceptions on the safe application of pesticides need to be improved. Hence, education on the scientific application of pesticides, which is limited in traditional education, needs to be strengthened [5].

References

[1] CNSB (China National Statistic Bureau) The production and growth rate of main industrial products (2011). http://www.stats.gov.cn/.

[2] World Bank World Development Indicators Database (2011). http://data.worldbank.org/data-catalog.

[3] Zhang F., Wang J., Zhang W. Nutrient use efficiencies of major cereal crops in China and measures for improvement. Chinese Journal of Soil Sciences 2008; 45(9): 915-924.

[4] Zhang W., Xu A., Ji H. Estimation of agricultural non-point source pollution in China and the alleviating strategies III: a review of policies and practices for agricultural non-point source pollution control in China, Chinese Journal of Agricultural Sciences 2004; 37(7): 1026-1033.

[5] Wei Q., Tao L.,Song X. Pesticide safety management in China: opportunity and development strategy, Chinese Journal of Quality and Safety Supervision 2011; 1: 11-14.

[6] Song X. Inventory of pesticide industry in 2010, Chinese Journal of Pesticide Science and Administration 2011; 32: 1-3.

[7] Gong Q., Zhang J., Li J. Analysis of factors affecting farmers' decision-making on fertilizer application, Chinese Journal of Agricultural Issues 2008; 10: 63-68.

[8] Gao H., Liang Z., Chen X. Analysis of factors influencing adaptation of the technology of formula fertilization by soil testing: based on the questionnaire survey of farmer households in Fujian Province, Chinese Journal of Fujian Agriculture and Forestry University (Philosophy and Social Sciences Edition) 2011; 14(1): 52-56.

[9] Han H. and Zhao L. Farmers' character and behavior of fertilizer application: evidence from a survey of Xinxiang County, Henan Province, China, Journal of Agricultural Sciences in China 2009; 8(10): 1238-1245.

[10] Deininger K., Feder G. Land Institutions and Land Markets, Handbook of Agricultural Economics, Volume 1A, Edited by B. Gardner and G. Rausser, Elsevier/North-Holland Press 2001: 289.

[11] Wang H., Xu X. Micro behaviors and the safety of agro-products: an analysis of rural production and resident consumption, Chinese Journal of Nanjing Agricultural University (Social Sciences Edition) 2004; 4(1): 23-28.

[12] Dai H. The impact analysis of agro-chain to farmer's safe production behavior, Chinese Journal of Hubei University of Economics 2010; 8(4): 73-78.

[13] Ma J. Analysis on amounts and determinants of fertilizer applied on cereal crops by the farmers: a case study of North China Plain, Chinese Journal of Agricultural and Technological Economics 2006; 6: 36-42.

[14] Liu Z., Sui X. Regional characteristics of fertilizer use in China, Chinese Journal of Resources Science 2008; 30(6): 822-828.

[15] Yang Z., Han H. Technical efficiency of fertilizer and influencing factors: Based on empirical analysis of wheat and corn, Chinese Journal of China Agricultural University 2011; 16(1): 140-147.

[16] Song Q., Fang J., Li Y. Discussion on Influencing Factors of Farming Household's Safety Agricultural Products Production, Chinese Journal of Chinese Agricultural Science Bulletin 2010; 26(24): 466-471.

[17] Li G., Zhu L., Ma L. Effects of nuisance-free agricultural products certification on the behavior of household pesticide use: a case study of Nanjing prefecture, Jiangsu Province, Chinese Journal of Rural Economics 2007; 5: 95-97.

[18] Zhu Y., Wu L. Comparison of pesticide application behaviors amongst farmers in different acreages, Chinese Journal of Zhejiang Agricultural Sciences 2010; 5: 1024-1029.

[19] Zhang Y., Ma J., Kong X., Zhu Y. Determinants on Farmers' use of green pesticides: empirical analysis on 15 counties of the Shanxi, Shaanxi and Shandong Provinces, Chinese Journal of Rural Economy 2004; 1: 44-52.

[20] Han W. Survey and analysis on environmental problems in the poverty rural areas of Sichuan Province, Chinese Journal of Rural Economics 2005; 11: 99-101.

[21] Zhang J., Li R. Rural environmental pollution and the countermeasures for sustainable development, Chinese Journal of Anhui Agricultural Sciences 2007; 35(15): 4588-4617.

[22] Jack J., John D. Econometric Methods (Fourth Edition). The McGraw-Hill Companies, Inc., New York 1997: 415-418.

[23] Seddighi H. R., Lawler K. A., Katos A. V. Econometric: A Practical Approach. Routledge Press, London 2000: 105-106.

[24] Bruin J. Newtest: Command To Compute New Test. UCLA: Academic Technology Services, Statistical Consulting Group 2006. http://www.ats.ucla.edu/stat/stata/ado/analysis/.

[25] Zheng X. Analysis of the influencing factors on the farmers' use of manures in Danjiangkou reservoir area, Chinese Journal of Hunan Agricultural University (Social Sciences) 2010; 11: 11-15.

[26] Yin C., Wu P., Zhang Y. Research on farmers' will to reduce the amount of crop fertilizer. Chinese Journal of Jiangsu Agricultural Sciences 2010; 1: 384-387.

[27] Sun Q. Analysis on the determinants of production of safety agricultural products by farms, Chinese Journal of Food and Nutrition in China 2008; 1: 15-17.

[28] Zhang W. Analysis on the market conditions of Folimat and some other highly toxic pesticides in China, Chinese Journal of Pesticides Marketing Bulletin 2008; 20: 27.

[29] Tao K., Zhou H. Furadan, folimat and some other pesticides are prohibited since tomorrow, Chinese Newspaper of Jiaxing Daily 2005; 6(30): 2.

[30] SCSC (Thirteenth Standing Committee of Suzhou People's Congress). Bulletin of supervision and management regulations on food safety in Suzhou Prefecture 2007. http://www.szfzb.gov.cn/news/fzb/2007/1/25/fzb-15-12-49-3707.shtml.

[31] Zhou W., Song X. Safeguarding the shopping baskets of the Capital through strengthened supervision and innovative services to: Interview with Zhang Lingjun, director of Beijing Pesticide Inspection Institute, Chinese Journal of Pesticide Science and Administration 2009; 30(12): 24-26.

[32] Li H., Fu X., Wu X. The wishes and influencing factors of farmers' safe use of pesticides: survey and analysis on 214 farmers of Guanghan Prefecture, Sichuan Province, China, Chinese Journal of Agricultural and Technological Economy 2007; 5: 99-104.

[33] Wooldridge J. M. Introductory Econometrics: A Modern Approach (2nd Edition). South-Western Thomson Learning, Mason 2003: 41, 81.

[34] Zhao J., Zhang Z. Analysis on the factors affecting safety agricultural production decisions of farms, Chinese Journal of Statistical Research 2007; 24(11): 90-92.

[35] Song X., Cao C., Cheng D., Zhang X., Zhang Z. The construction of security system of agricultural products and controlling system of pesticide pollutions in Beijing, Chinese Journal of Agricultural Environment and Development 2008; 6: 52-55.

[36] Bo X. Shanghai locally grown vegetables and completely safe, Chinese Newspaper of Wenhui Daily 2009; 10(24): 2.

[37] Gu D. Shanghai pesticide industry takes on a dumbbell-shaped developing state, Chinese Newspaper of AgriGoods Herald 2004; 4(25): 17.

Chapter 5

Farmer Perceptions on the Safety of Agricultural Products

Xiaoou Gao and Min Song

5.1 Introduction

With the accelerating process of global integration and the increasingly fierce competition in the international market, the safety of agricultural products, as the most basic material for human survival and development, has come into global focus. China is a large agricultural country, in which the output and the export of agricultural products have constantly increased over recent years. In 2011, the agricultural output in China was 737.113 billion US dollars, accounting for 17.3% of global agricultural output, which is far ahead of other countries in the world. Since 2008, China has become the fifth largest exporter of agricultural products [1]. Statistics show the import and export of agricultural products amounted to 154.03 billion US dollars in 2011 and the export volume was 60.13 billion US dollars. However, China is facing greater challenges due to the outbreak of a series of agro-food safety scandals in recent years.

With the development of industry and the economy, promoted by incentive polices, chemical products have been increasingly put into agriculture in China over recent decades. China has become the largest user of fertilizers, pesticides and plastic film in the world [3]. In 2010, Chinese agriculture consumed 55.61 million tons of chemical fertilizer, 1.75 million tons of pesticide and 2.17 million tons of plastic film [4-5] - all far higher than the world average. Nearly 60-70% of the chemical fertilizer, 60% of the pesticide and 40% of the plastic film are exposed to the environment directly [6-7], posing ultimate threats to the safety of agro-products, food and human health. Recently, multiple agro-food safety incidents occurred, e.g., the Qingdao drug leak incident of 2010, the Hainan drug cowpea incident of 2011 and the Guangdong drug watermelon incident of 2007. Disclosure of these agro-food safety scandals has led to a series of severe consequences. First of all, considerable outbreaks of food borne disease have fundamentally undermined public trust. According to the Report on Chinese Food Safety in 2011-2012 released by the Chinese magazine of *Well-off* and Tsinghua University, it was reported that 63.7% of the respondents believed that food safety in China is bad and that 80.4% of the respondents thought food is not safe at all in China. Secondly, food safety

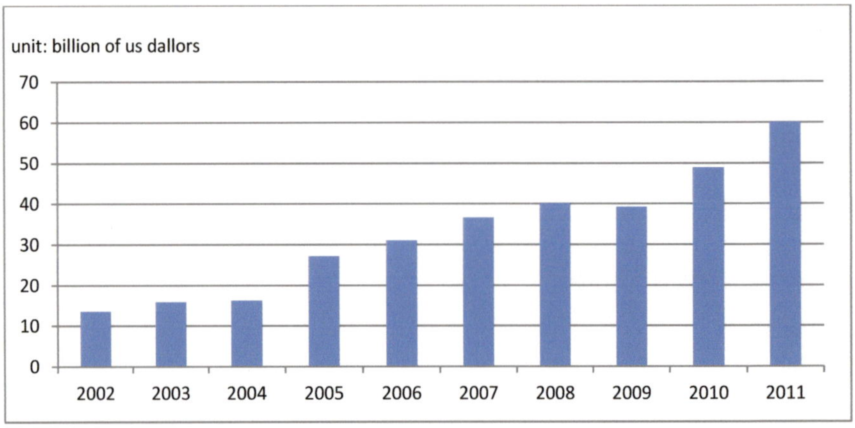

Source: China Ministry of Agriculture [2]

Figure 5.1 Volume of exported agro-products in China

scandals have serious negative impacts on food exports. On the one hand, they undermine the international competitiveness and market reputation of food exported from China. Massive exposure of food safety incidents in China greatly reduces foreign consumers' confidence in food labelled as made in China. A survey carried out in Korea shows that nearly 90% of respondents think food imported from China is not safe [8]. On the other hand, the international market will raise barriers to limit food imported from China. For example, a number of countries applied stricter standards on products made in China after the disclosure of the poisoned capsule event in early 2012. Thirdly, food safety accidents often result in significant economic losses. Take the Sanlu incident as an example, it led to the bankruptcy of a famous enterprise and brand, simultaneously causing the entire milk industry to suffer. During the incident, a vast amount of milk was poured away, and a great number of dairy cows were slaughtered.

In the previous literature, more and more economists have studied food safety issues using empirical analysis methods, especially on the topics of consumer behaviours, awareness and willingness to pay for certified safe food. For example, Chern, W. S., et al (2002) [9] studied consumers' willingness to pay for genetically modified vegetable oils through carrying out a survey in Shikoku of Japan, Norway, Taiwan and the United States. Georges G, et al (2006) [10] selected interviewed consumers from 12 European countries, divided them into four groups, and managed to understand their perception on the food traceability system. Wang (2003) [11] carried out a survey of 289 consumers in Tianjin, China, and analysed their decision process and characteristics when choosing safe food. Through a survey in Zhejiang Province, China, Zhou (2004) [12] revealed that consumers are concerned about vegetable safety - illustrating their negative attitudes about the current situation. They are very willing to pay the extra cost for certified safe vegetables, but the price of the safe vegetable should be no more 10% to 20% higher than the ordinary vegetable. Zeng et al (2008) [13] studied consumers' willingness to pay for moon cake with safe additives, through a survey of 396 consumers sampled from 25 supermarkets in Beijing. But at least until now, studies on food safety from the farmers' perspective are extremely rare.

The objective of this chapter is to understand the situation of agro-food in China through analysing farmers' confidence and determinants on their own agro-products safety. Different from consumers, farmers are both managers and producers of agro-products, and they possess much information about agricultural production - from sowing, fertilizing and controlling pests to harvesting. Therefore, their confidence on the safety of their agro-products provides a new perspective to studying the real quality of agro-products and ensuring food safety.

The remaining sections will be organized as follows: Section 2 represents the field survey and characteristics of the sampled farmers, including their demographic information, farming condition and knowledge of agro-products' quality. Section 3 describes farmers' confidence on their own agro-products and analyses impacting factors by using the binary logit regression model. In Section 4, some conclusions and recommendations are proposed.

5.2 Data source and demographic characteristics

5.2.1 Data source

As described in the Chapter 3, the data in this chapter is obtained through a field survey of 560 samples from 21 villages carried out in January to March 2011, which covered six provinces of eastern China (Figure 3.1). In addition to the number of sampled farmers among the six provinces, their distribution between the north and south, the two metropolises and other areas are also included. In this research, the main research objective is to analyse farmers' confidence and determinants on their own agro-products safety; there are 346 respondents available to apply to this research, therefore the sample size for the current research is 346 (Table 5.1).

Region	Beijing	Hebei	Shandong	Shanghai	Jiangsu	Zhejiang
Number	66	85	69	64	57	63
Location	North=1 (214); South=0(132)					
City scale	Metropolises=1 (55); Not Metropolises =0 (291)					
Source: field survey by the authors in 2011						

Table 5.1 Distribution of sampled farmers

5.2.2 Demographic characteristics

The results of this survey show that farmers' average age is 48.92 years. Most of them are middle-aged and aged, a half of them are older than 45 years old and younger than 60 years old, while 17.34% of them are older than 60 years old. The number of male farmers is much greater than that of female, accounting for 84.10% in total. Farmers' general education level is low, as 80.92% of them are educated to no higher than high school. There are 98 farmers who participated in off farming works in 2010, accounting for 28.32 in total. From the perspective of household income, 16.23% of the farmers' household income in 2010 is less than 10,000 yuan, 11.30% of the farmers' household income is more than 50,000 yuan, and 72.47% of the farmers' household income is between 10000 yuan and 50000 yuan (Table 5.2).

Characteristics	Count	Unit	Mean	Min	Max	Std. D
Age	**346**	**years**	**48.92**	**21**	**85**	**11.26**
Gender	346	dummy	Male=1 (84.10%); Female=0 (15.90%)			
Education	346	dummy	Illiteracy=1 (4.05%); Primary=2 (28.03%); Middle=3 (48.84%); High=4 (14.74%); Advanced=5 (4.34%)			
Off farming work	346	dummy	Do off farming work=1 (28.32%), Not do off farming work=0 (71.68%)			
Income	346	dummy	Annual income of a household Less than 10,000 yuan=1 (16.23%), 10,000-30,000 yuan=2 (39.13%), 30,000-50,000 yuan=3 (33.33%), more than 50,000 yuan=4 (11.30%)			

Source: field survey by the authors in 2011

Table 5.2 Demographic characteristics of farmers in 2010

5.2.3 Farming condition

As shown in Table 5.3, sowing scales of grain crops are quite different among sample farmers, where the maximum sowing area is 112 mu, much more than the minimum sowing area of 0.2 mu. More than 80% of farmers sell their agro-products. In agricultural production, a great number of farmers used chemical fertilizer by spreading it on the surface of the land, while 35.55% of them used manure as an auxiliary fertilizer. Some farmers still used prohibited highly toxic pesticides, such as dichlorvos (DDVP). Meanwhile, very few farmers responded as having adopted biological methods to control pests.

Characteristics	Count	Unit	Mean	Min	Max	Std. D
Sowing area	346	mu	5.20	0.2	112	7.40
Selling ratio	346	%	61.78	0	100	133.89
Manure using	346	dummy	Using manure = 1 (35.55%); Not using manure = 0 (64.45%)			

Source: field survey by the authors in 2011

Table 5.3 Farming condition of the sampled farmers

5.2.4 Farmers' perception on the safety of agro-products

In the farming behaviours section of our questionnaire, there are two questions about agro-product safety. The first one deals with what is the prior determinant for farmers when using fertilizers and pesticides. The result shows that 71.47% of farmers are most concerned about yield, 17.58% of farmers are most concerned about cost, and 7.49% of farmers are most concerned about the quality of agricultural products. The second question deals with what farmers think are the consequences of overdosing. As a multiple-choice question, farmers choose the answer of imperilling food safety which accounts for an overwhelming ratio of 71.10%, followed by another choice of imperilling sprayers' health at 51.73%. Next is the choice of environment pollution at 50.58% (Table 5.4).

Characteristics	Count	Unit	description
Determinant of using fertilizer and pesticides	346	dummy	Food quality =1 (7.49%); Others = 0 (92.51%)
Consequence of overdose	346	dummy	Food quality =1 (71.10%); others = 0

Source: field survey by the authors in 2011

Table 5.4 Farmers' perception on agro-product safety

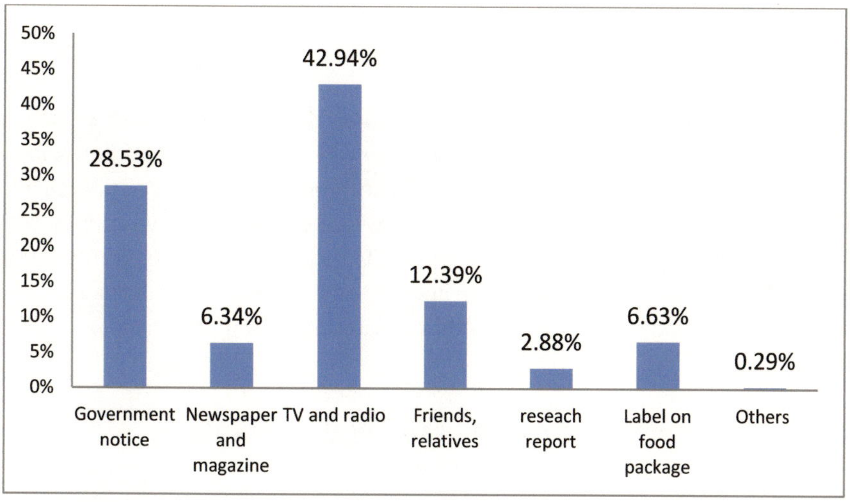

Source: field survey by the authors in 2011

Figure 5.2 Farmers' information channels

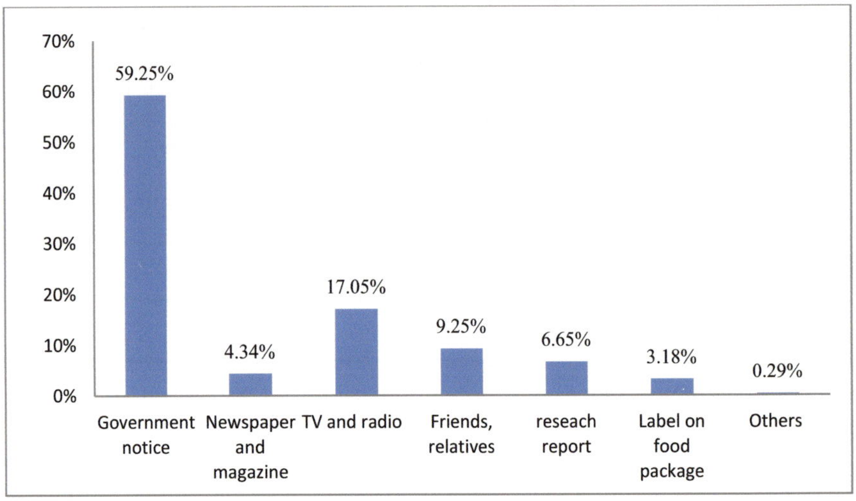

Source: field survey by the authors in 2011

Figure 5.3 Farmers' reliance on different information sources

As to the farmers' conception on agricultural pollution, the data shows that TV and radio are their most common ways of obtaining food safety information and related news, followed by government notices, friends and relatives, labels on food packages, newspapers and magazines, research reports and others (Figure 5.2). However, they have the mostly reliance on information provided by the government, while TV and radio are ranked in second palace, with labels on food packages ranked last. In another words, most of the farmers have no trust at all in the information disclosed by the food manufacturers (Figure 5.3).

5.3 Analysis on farmers' confidence and determinants

5.3.1 Farmers' confidence on the safety of agro-products

As shown in Figure 5.4, as to farmers' confidence on safety of their own agro-products, 19.65% of them think that they are very safe, 63.78% of them think that they are safe, 11.27% of them think that they are not safe, while 1.16% of them think that they are not safe at all, and 4.05% of them responded as being neutral.

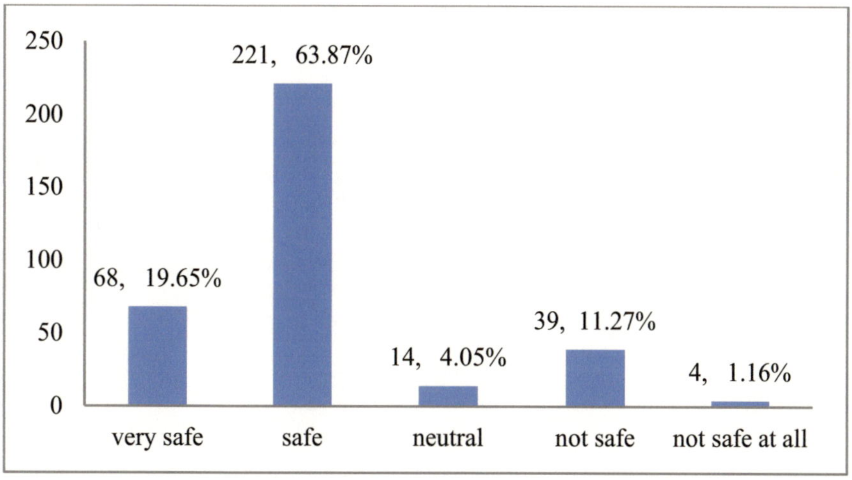

Source: field survey by the authors

Figure 5.4 Farmers' confidence on agro-product safety

5.3.2 Factors affecting farmers' confidence

As to the perception on agro-product safety, farmers' confidence is influenced by their demographic characteristics [14], farming condition and knowledge on agro-production safety [15]. In our study, gender, age, education level, the doing off-farming work or not and household income are included as demographic factors. The sowing area of field crop, selling ratio of agro-production and use of manure or not are included as farming condition factors. Answers to the questions about agro-products safety are included as farming condition factors. In addition, location and city scale are included to examine the significance of geography.

5.3.3 Model specification

A binary logit regression model is adopted to examine factors impacting on farmers' confidence, which is shown as:

$$y = \beta_0 + \beta_1 x_1 + \beta_2 x_2 + \ldots + \beta_k x_k + u \qquad (5\text{-}1)$$

while the distribution function of y is:

$$f(y) = P^y (1-P)^{1-y}, \ (y = 0, 1) \qquad (5\text{-}2)$$

where when farmers are very confident or rather confident, y=1; otherwise, y=0; X (x_1, x_2, …, x_k) represents a series variables affecting farmers' confidence on agricultural production safety; u is the random error. The model is calculated by using SPSS 18.0 for Windows.

Characteristics	B	S.E.	Wald	df	Sig.	odds ratio
Location**	0.989	0.440	5.056	1	0.025	2.688
City	0.218	0.423	0.267	1	0.606	1.244
Gender	-0.137	0.460	0.089	1	0.765	0.872
Age	-0.019	0.016	1.344	1	0.246	0.981
Education level	-0.240	0.220	1.187	1	0.276	0.787
Off-farming work	-0.315	0.351	0.803	1	0.370	0.730
Income	-0.205	0.223	0.839	1	0.360	0.815
Sowing area*	-0.054	0.033	2.800	1	0.094	0.947
Selling ratio	0.007	0.005	2.197	1	0.138	1.007
Manure using***	1.062	0.408	6.773	1	0.009	2.892
Determinants of using fertilizer and pesticide	0.016	0.050	0.096	1	0.756	1.016
Consequence of overdose	0.447	0.354	1.588	1	0.208	1.563
Omnibus tests of model coefficients	Chi-square (12) = 25.948, Sig. =0.010***					

Note: ***, **and *represent statistical significance in the level of 1%, 5% and 10%, respectively.

Software: SPSS 18.0

Table 5.5 Effects of characteristics on farmers' confidence

5.3.4 Result and discussion

As shown in Table 5.5, only using manure or not, location and sowing area are significant, while other factors are measured as being insignificant upon farmers' confidence on the safety of their own agro-products.

As manure is considered less harmful to agricultural environments and products, farmers using manure are more confident about their own products. In the north there are 86 farmers who use manure, accounting for 40.19% of northern farmers. Meanwhile, there are 37 farmers using it in the south, 28.03% of the sampled farmers from the south. At the same time, the data suggests that southern farmers focus more on the safety of agro-products, and they have more knowledge and better understanding of the harmful effects of agricultural chemicals. Therefore, southern farmers' attitudes on agro-product safety are more negative. Furthermore, farmers who own larger scales of land are likely to be professional. They not only have more knowledge on agro-product safety, but also tend to pursue more yields through applying more agro-chemicals. As a result, they lack confident on the safety of their own agro-products.

5.4 Conclusion and recommendation

5.4.1 Major conclusions

Farmers do not have sufficient awareness of agro-product safety during production. When using chemical fertilizers and pesticides, only 7.51% of the farmers considered the negative impacts on the safety of agro-products, while most of the farmers preferred to concentrate on increasing the yields.

Farmers' knowledge of agro-product safety is not very comprehensive. On the one hand, influenced by the publicity surrounding pesticide residues incidents, 71.10% of them think overdosing will influence agro-products' safety. However, 58.38% of farmers understand well or better understand the withdrawal period. On the other hand, farmers using manure are very confident about the safety of their agro-products. In fact, the safety of manure is not perfect, despite the fact that it is less harmful to agricultural environments and products than agro-chemicals.

Farmers who sow more farmland are less confident on their products, which may be because the farmers who own more land are more likely to be professional farmers. They tend to pursue productivity through inputting more agro-chemicals.

5.4.2 Policy recommendations

To improve farmers' awareness on agro-product safety, more related information needs to be released by the mass media including TV, radio, etc. As revealed by the analyses in this chapter, the public media are the most commonly used channels by which the farmers gather information and the recommendations above must be effective in helping them to improve their behaviours.

This survey reveals many farmers have poor knowledge about the withdrawal period and outcome of overusing agro-chemicals. Hence, concerning public education on the withdrawal period and proper use of chemical fertilizers and pesticides, it needs to be strengthened.

Strengthening the supervision of farmers, especially farmers who own large-scale land, is also necessary. This is because, as professional farmers, they are concerned with agro-products yields rather than quality. Therefore, related government departments should strengthen their supervision of farming behaviours.

There are some limitations of this study, e.g., in the model, amounts of fertilizers and pesticides should be included as variables and location may not be significant in influencing farmers' confidence on agro-product safety. However, the data collected is not complete. In future studies, more scientific variables should be included, with the adoption of further optimized models.

References

[1] Wu, T. China's agricultural export doubled in a decade after accession to the WTO (2011). http://www.cfqn.com.cn/Article/2011/2093q/2093a/19090786589501.htm.

[2] China Ministry of Agriculture. China Statistical Yearbook 2012. Shenyang: Liaoning Education Press, 2012.5.

[3] Liu G P, et al. Current situation and measures of agricultural pollution in China. Studies in International Technology and Economy, 2006; 9(4):17-21.

[4] CNSB (China National Statistical Bureau). Using of agricultural plastic film and pesticide in different regions in China, http://www.stats.gov.cn/.

[5] CNSB (China National Statistical Bureau).Using of chemical fertilizers in different regions in China, http://www.stats.gov.cn/.

[6] Sun J L. Review of agricultural pollution and preventive technology in China. Journal of Jishou University (Natural Science Edition) 2008; 29(5):99-128.

[7] Hou B, Hou J, Wang Z W. Farmers' perception on pesticide residue and its influence on pesticide application. Heilongjiang Agricultural Sciences 2010; (2):99-103.

[8] The survey shows that nearly 90% of Koreans distrust China's food safety http://gb.cri.cn/27824/2012/08/16/6071s3812468.htm.

[9] Chern, W.S., Rickertsen, K., Tsuboi, N., Fu T. Consumer acceptance and willingness to pay for genetically modified vegetable oil and salmon: A multiple-country assessment. AgBioForum, 2002; 5(3): 105-112. From: http://www.agbioforum.org.

[10] Georges G, Rafia H. Consumers' Perception on Food Traceability in Europe. International Food & Agribusiness Management Association World Food & Agribusiness Symposium, Buenos Aires, Argentina, 2006; (7):1-11.

[11] Wang Z. Perception on food safety and consumption decides: empirical analysis of individual consumers in Tianjin. Chinese Rural Economy 2003; 4:41-48.

[12] Zhou J. Analysis on consumers' attitude, perception and purchasing behaviors on vegetables safety: basis on a survey in Zhejiang Province. Chinese Rural Economy 2004; (11):44-52.

[13] Zeng Y., Liu Y., Wang X. Analysis on willingness to pay for food safety with Hierarchical model: in the case of consumers' willingness to pay for moon cake additives. Journal of Agro-technical Economics 2008; 1:84-90.

[14] Feng Z., Li Q. Analysis on farmers' perception and determinants of agro-products safety. Agricultural Economic Problems 2007; (11):22-26.

[15] Zhou J. Analysis on farmers' behaviors and determinants of vegetable quality controlling: evidence from 369 vegetable farmers in Zhejiang Province, Rural Economy of China 2006; (11):25-34.

Chapter 6

Farmer Perceptions on Risk Sources and Management

Hui Zhou and Teruaki Nanseki

6.1 Introduction

Agricultural management comprises several parts, including risk management, information management, technology management, etc. Understanding and estimating potential risk quickly and properly is very important. This is a first step in risk management. Additionally, collecting, processing, providing and analysing information effectively is also necessary in risk management as well as in agricultural management. This is a first step in information management. In information management, new technologies such as information and communication technology (ICT) have increasingly been used on management strategies (Figure 6.1).

In this chapter, farmers' perceptions of risk and their points of view on risk management strategies are studied. China is the third largest milk producer in the world, right after the USA and India [2]. As a way to increase farmers' income and improve people's diet, the Chinese government encourages farmers to raise more cows and produce more milk. However, risks in both food and agriculture including food contamination by chemicals, animal diseases and price variability in China lead to threats, both to dairy farmers and consumers. Dairy farmers are most negatively affected by these risks, as they belong to the weakest part of the whole milk supply chain. It is important for policy makers to understand farmers' perceptions and their responses to the risk in order to help dairy farmers reduce their losses. In that, dairy farmers are used as an example, and dairy farmers' risks and risk management strategies are mainly examined in this research.

Risk is uncertainty that affects an individual's welfare, and is often associated with adversity and loss [3]. In response to risky situations, farmers should be involved in risk management, making choices among alternatives so as to reduce the effects of the risks. The main research objectives are: to examine the dairy farmers' perception of risk and to examine the risk management strategies of dairy farmers.

Source: Nanseki (2011) [1]

Figure 6.1 Full picture of management in agriculture

6.2 Literature review

Agriculture is inherently risky. Farm outputs depend on weather and biological processes over which producers have little control, and competition in domestic as well as international markets exposes agricultural producers to unanticipated price fluctuations [4]. However, some of the risks in the livestock sector are different from those in the crop sector. This chapter mainly focuses on dairy farmers' risk perception and risk management strategies.

There are several studies about farmers' risk perception. The Economic Research Service (ERS) has summarized American studies [3]. In the US, dairy farmers were most concerned about commodity price risk, production risk and changes in government laws and regulations. Dairy farmers in Arizona perceived the costs of operating inputs to be the greatest source of risk [5]. Dairy farmers in New Zealand viewed price risk and rainfall variability as the highest risks [6]. Meuwissen et al (2001) [7] found that Dutch livestock farmers considered price and production risks to be most important. In Japan, the biggest and the sometime occurring risks to livestock farmers are decrease of production by animal death and decrease of quality by equipment breakdown [1].

ERS also found that keeping cash at hand was the chief risk management strategy for every farm size, for every commodity specialty, and in every region studied; use of derivative and insurance markets was also considered important [3]. Maintaining animal health was viewed as the most effective strategy. The research in Holland found that, producing at the lowest possible costs and insurance were the most important risk management strategies [7]. A study among Finnish farmers found changes in agricultural policy to be the most important risk factor, while maintaining adequate liquidity and solidity was the most important management response [8].

In China, there are several risks that the whole dairy industry faces. The food safety problem is considered as one of the biggest risk for the dairy industry [9-10]. Lack of a complete law and regulation system, having a breed of cow which is not pure and low yield are considered important risks for the whole industry [9]. However, in China, most studies in this field focused on the macro level - from the point of view of the whole industry - and there are very few studies focusing on dairy farmers' perceptions presently. Therefore, this research focuses on uncovering dairy farmers' perceptions of risk and their risk management strategies in China.

6.3 Data and method

6.3.1 Questionnaire and analysis technique

In this study, a survey has been conducted to sample farmers in order to ascertain the risks they are most concerned about and their risk management strategies to cope with such risks. There are a number of sources of risk and risk management strategies listed in the questionnaire. Based on previous studies of ERS (1997) [3], Wilson et al. (1988) [5], Meuwissen et al. (2001) [7] and Abdi (2004) [11], 22 sources of risk and 18 risk management strategies were given in the questionnaire at first. However, after consulting with experts and considering the current situation in China, finally, 19 sources of risk and 15 risk management strategies were selected in the questionnaire using in this study (Table 6.1 and Table 6.2). In China, since 2006 farmers do not need to pay tax, so that all the risks associated with tax payment are deleted.

To understand which risk and risk management strategies have more weight, a 5-level Likert scale is used. Dairy farmers are requested to judge and label different levels of each risk and risk management strategies. Level 1 stands for the least important and level 5 stands for the most important.

Farmers' risk perception and risk management strategies were studied by descriptive analyses. When the number of variables is large, a multivariate analysis technique such as principal component analysis (PCA) is effective to reduce its dimensionality. PCA is one multivariate technique that analyses a data table in which observations are described by several inter-correlated quantitative dependent variables [12]. Its goal is to extract the important information from the data, to represent it as a set of new orthogonal variables called principal components, and to display the pattern of similarity of the observations and of the variables as points in maps. In this research, therefore PCA is used to analyse the survey data. PCA is suitable to extract the principal risk and strategy from many risks and risk management strategies. To reduce the dimensionality of the variables, PCA is a powerful tool to do such work, without losing much information. Additionally, this method is often used in similar pieces of research.

PCA is a proper way to analyse data and is adopted in this case. All analyses including PCA were carried out by using SPSS 13.0 for Windows [13].

Category	Level
1. Changes in consumer preferences	1 2 3 4 5
2. Production diseases such as mastitis	1 2 3 4 5
3. Domestic epidemic animal diseases such as para tuberculosis	1 2 3 4 5
4. Non-domestic epidemic animal diseases such as foot and mouth disease	1 2 3 4 5
5. Misuse of veterinary drugs and veterinary drug residues	1 2 3 4 5
6. Related food safety issues occurring	1 2 3 4 5
7. Food safety news in media	1 2 3 4 5
8. Milk yield variability	1 2 3 4 5
9. Milk price variability	1 2 3 4 5
10. Corn yield variability	1 2 3 4 5
11. Corn price variability	1 2 3 4 5
12. Crop yield variability	1 2 3 4 5
13. Crop price variability	1 2 3 4 5
14. Costs of operating inputs	1 2 3 4 5
15. Changes in technology	1 2 3 4 5
16. Changes in government support payments	1 2 3 4 5
17. Health problems among family members	1 2 3 4 5
18. Hard to get loan	1 2 3 4 5
19. Fire damage, flood, drought, or other damage	1 2 3 4 5

Note: 1=Least Important, 5=Most Important

Table 6.1 Risk sources as categorized by dairy farmers

Category	Level
1. Production contracts	1 2 3 4 5
2. Liquidity - keeping cash at hand	1 2 3 4 5
3. Using consultant service or consultant extension workers	1 2 3 4 5
4. Joining the extension training programme	1 2 3 4 5
5. Asset flexibility - farm building with multiple uses	1 2 3 4 5
6. Keeping fixed costs low - rent machinery rather than purchase it	1 2 3 4 5
7. Shared ownership of equipment, joint operations	1 2 3 4 5
8. Off-farm work	1 2 3 4 5
9. Collecting information	1 2 3 4 5
10. Risk reducing technologies	1 2 3 4 5
11. Joining the farmers' corporation	1 2 3 4 5
12. Producing at lowest possible cost	1 2 3 4 5
13. Prevent/ reduce livestock diseases	1 2 3 4 5
14. Buying agricultural insurance	1 2 3 4 5
15. Slaughter or sell the cows, quit the business	1 2 3 4 5

Note: 1=Least Important, 5=Most Important

Note: keeping cash at hand is considered as one of the risk management strategies that can handle all kinds of risk and fits all sizes of farms; producing at the lowest cost means dairy farmers will do everything that could help them to keep the variable costs low, such as use the least amount of feed; preventing and reducing livestock diseases here means during the dairy feeding, farmers add some medicine to the feed to avoid certain diseases; using human medicine instead of animal medicine and so on.

Table 6.2 Dairy farmers' risk management strategies

After PCA, cluster analysis-hierarchical clustering was used to separate the farmers into different groups. Cluster analysis or clustering is the task of assigning a set of objects into groups (called clusters) so that the objects in the same cluster are more similar (in some sense or another) to each other than to those in other clusters.

Cluster analysis itself is not one specific algorithm, but the general task to be solved. It can be achieved by various algorithms that differ significantly in their notion of what constitutes a cluster and how to efficiently find them. Popular notions of clusters include groups with low distances among the cluster members, dense areas of the data space, intervals or particular statistical distributions.

6.3.2 Study site and respondent attributes

The field survey was carried out in Inner Mongolia and Hebei Province in April and June 2010, respectively. A sample of 168 dairy farmers is available for analysis in this study. Both Hebei Province and Inner Mongolia Autonomous Region are the main grain-growing and important milk production areas in China. The borders of the two provinces are adjacent to each other and located in the north part of China. Inner Mongolia is the largest milk producer in China and Hebei Province is the third biggest milk producer in China (Figure 6.2). By the end of 2010, the total population of the two regions amounted to 24.72 million and 71.94 million, accounting for 1.84% and 5.37% of the national population. With regard to average disposable income per urban household, both Inner Mongolia (17.70 thousand yuan) and Hebei Province (16.26 thousand yuan) are lower than the national average level of 19.11 thousand yuan, ranking the 10th and 14th among 31 mainland provincial regions of China. Meanwhile, the average consumption expenditure per urban household in Inner Mongolia and Hebei were 13.99 and 10.32 thousand yuan, while the national mean was 13.47 in the same time period [14]. Thus, the two regions have sound representativeness with regard to the consuming capability of China and are suitable as the survey areas.

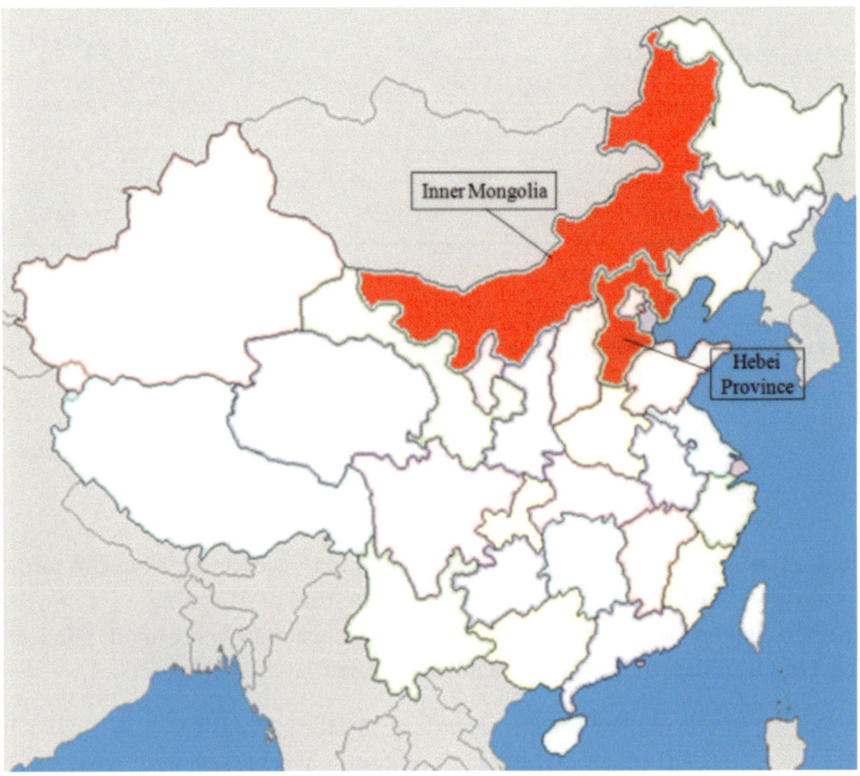

Source: revised based on http://www.chinamapxl.com/

Figure 6.2 Location of Inner Mongolia and Hebei Province

In this survey, most respondents are male, and nearly 80% of the respondents obtained less than nine years of education (Table 6.3). Most of the dairy farmers own less than 20 cattle, more than half of the dairy farmers only own around 10 cattle in each household. Meanwhile most dairy farmers have their own lands to grow corn which is used as the main feed for cows. Compared to countries such as the USA and some European countries, the annual average yield of each cow is not as good as could be hoped for, being only around 4000 kg. In the USA, a pure Holstein cow can produce around 10000 kg of milk per year. This is because the cows in the study area are pure Holstein mixed with local cattle, meaning the yield of second generation mixed Holstein cows is lower than pure Holstein cows.

Category			**Category**		
Gender	Male	69%	Education	Illiterate	13%
	Female	31%		Primary School	26%
Average yields of each cattle		3952 kg		Junior High school	41%
				Senior High school	17%
Average Age		42 year		College	3%

Source: consumer-survey 2010

Samples=168

Table 6.3 Demographic features of sample farmers

6.4 Result and discussion

6.4.1 Farmers' perception of risk

Table 6.4 shows the dairy farmers' perception of risk in both Inner Mongolia and Hebei Province. The table shows that, on average, the highest scores are given to variability of milk price in both places. In general, price and production risks are considered as the biggest risk. In Inner Mongolia, milk price variability, non-domestic epidemic animal diseases, such as foot and mouth disease, and corn price variability are considered as the three biggest risks. However, the situation in Hebei Province is slightly different. Besides milk price variability and non-domestic epidemic animal diseases such as foot and mouth disease, other food safety issues are also considered important risks for dairy farmers. In 2008, the milk powder scandal begun in Hebei Province at first, and the scandal finally made the local dairy company bankrupt. It subsequently became difficult for some local farmers to sell fresh milk. This shows that the direct experience of a milk and food safety issue in past still negatively affects dairy farmers in Hebei Province.

Category	Total		Hebei Province		Inner Mongolia	
	Mean	Rank	Mean	Rank	Mean	Rank
Milk price variability	4.86	1	4.89	1	4.82	1
Food safety issue news in media	4.47	2	4.78	2	4.01	4
Non-domestic epidemic animal diseases such as foot and mouth disease	4.42	3	4.31	4	4.57	2
Related food safety issues occurring	4.41	4	4.74	3	3.93	6
Corn price variability	4.15	5	4.19	5	4.09	3
Production diseases such as mastitis	3.83	6	3.82	7	3.84	7
Misuse of veterinary drugs and veterinary drug residues	3.70	7	3.67	9	3.74	8
Crop price variability	3.67	8	4.04	6	3.22	11
Corn yield variability	3.65	9	3.41	10	3.99	5
Crop yield variability	3.31	10	3.37	11	3.12	13
Domestic epidemic animal diseases such as Para tuberculosis	3.30	11	3.11	12	3.59	9
Fire damage, flood, dry, or other damage	3.23	12	3.76	8	2.46	18
Health problems among family members	3.21	13	3.07	13	3.41	10
Milk yield variability	3.10	14	3.06	14	3.16	12
Costs of operating inputs	2.62	15	2.38	15	2.97	15
Changes in government support payments	2.58	16	2.31	16	2.99	14
Changes in technology	2.38	17	2.28	17	2.53	16
Changes in consumer preferences	2.36	18	2.26	18	2.51	17
Hard to get loan	1.73	19	1.61	19	1.91	19

Source: calculation based on survey 2010

Note: the order of risk is based on mean score of each one (Column 2)

Table 6.4 Mean score and rank for source of risk

These results in China are similar to those obtain in related research carried out in some other countries, such as price variability in products, animal diseases such as foot and mouth disease and others. However, in those developed countries, mastitis is not such a serious disease to cows, but in China, it is still a big problem which significantly affects Chinese dairy farmers. This is because in most developed countries, better and more modern methods are used in raising cows, while in China old-fashioned and traditional ways still predominate among farmers allowing this kind of production disease to occur more easily. For example, in most developed countries, the total mixed ration (TMR) technique is used to raise cows. However, those small scale dairy farmers are unaware of TMR in China. Inappropriate ways of raising cows may cause mastitis and other diseases. In most developed countries, only machines are used to milk cows, but in China, some farmers still use hands to milk cows. Additionally, the disinfection technique used on the cows in China is not good enough, helping to aggravate the problem of mastitis for Chinese dairy farmers. Meanwhile, because of this kind of disease, large amounts of antibiotics are used in China, and this causes antibiotic residues to occur more often than in developed countries.

Table 6.5 shows the Varimax rotated factor loadings for source of risk. Before starting PCA, sampling adequacy was checked to detect if the data will factor well. In SPSS, sampling adequacy is measured by using the Kaiser-Meyer-Olkin (KMO) criterion. KMO varies from 0 to 1 and a KMO higher than 0.5 implies that PCA can be applied to the data [15]. The KMO value of source of risk is 0.517 and is acceptable for PCA. The number of risk was reduced by applying PCA. This resulted in seven factors with eigenvalues larger than 1 and in total accounted for 77.4% (which can be regarded as satisfactory in social sciences).

In addition, Table 6.5 shows the factor loadings for the source of risks. According to the loadings, the factors can be described as 'production risk', 'institutional risk', 'animal disease', 'input market risk', 'milk contamination risk' and 'personal risk' and 'output market risk', respectively.

On the first factor 'production risk', high loading goes to input especially feed variability and some changes in farm operating input. However, those risks from outside of the farm business, such as fire damage, flood damage and other food safety issues, show a high negative loading. With regard to food safety issues occurring, this is also included in production risk. It also can be considered as some risk outside of the dairy business itself, so that the loading is negative but not very high. High loading of government support policy change and changes in technology are related to the second factor 'institutional risk'. In China, most technology improvements are done by national research institutes. Change in technology is also considered as an institutional risk. Animal diseases, such as foot and mouth disease, mastitis and para tuberculosis, veterinary residue and misuse of veterinary medicine also related to the third factor 'animal disease risk'. However, the government provides the vaccine for foot mouth disease twice a year for farmers free of charge, so mouth and foot disease shows a high negative loading.

On the other hand, mastitis and para tuberculosis are not included in the government free vaccine programme. Feed price, such as corn and other crop prices, show a high loading and contributes to the fourth factor of 'input market risk'. The factor 'milk contamination risk' mainly explains the situation in Hebei Province. The factor 'personal risk' is likely to reflect changes in the family, only health problems among household members shows a high loading in this factor. In addition, the milk price variability and change in consumer preferences are related to the last factor 'output market risk'. 'Changes in consumer preference' refers to the situation where, for example, consumers choose to consume more soybean milk instead of cow milk. However, milk yield is negatively correlated with milk price, so it shows a negative sign in this factor.

Category	Component						
	Product-ion	**Institution**	**Animal disease**	**Input mar-ket**	**Milk contami -nation**	**Pers onal**	**Output Market**
Crop yield variability	**0.834**	0.286	-0.019	0.102	-0.029	-0.004	-0.012
Costs of operating inputs	**0.810**	0.347	-0.088	0.218	0.247	0.061	-0.009
Fire, flood, drought, or other damages	**-0.796**	0.057	-0.029	-0.055	-0.118	-0.151	0.008
Corn yield variability	**0.715**	0.260	-0.182	-0.09	-0.083	-0.139	-0.426
Related food safety issues occurring	**-0.507**	0.208	0.436	0.173	-0.426	-0.218	0.123
Changes in government support payments	-0.112	**0.888**	0.060	-0.218	0.043	-0.086	-0.038
Changes in technology	0.406	**0.578**	0.362	0.032	-0.079	0.400	-0.033
Domestic epidemic diseases such as para tuberculosis	0.080	0.101	**0.866**	-0.078	0.135	-0.058	-0.094
Misuse of veterinary drugs and veterinary residues	-0.014	-0.212	**0.800**	0.276	0.140	0.167	0.068
Non-domestic epidemic animal diseases such as foot and mouth disease	0.297	0.242	**-0.725**	0.008	-0.181	0.223	-0.179
Production diseases such as mastitis	0.012	-0.017	**0.644**	-0.242	0.04	-0.146	-0.118
Corn price variability	0.186	0.000	0.084	0.785	0.059	-0.243	0.197
Crop price variability	-0.396	0.325	0.012	**0.724**	-0.220	-0.061	0.034
Hard to get load	0.133	0.372	0.298	**-0.559**	0.373	0.153	0.332
Food safety issue news in media	-0.479	0.176	0.443	0.111	**-0.553**	-0.092	0.149
Health problems among family members	0.044	0.061	-0.482	-0.092	0.147	**0.692**	0.210
Changes in consumers preferences	0.061	-0.161	0.096	-0.209	-0.063	-0.129	**0.799**
Milk yield variability	0.093	0.263	0.171	0.321	0.414	0.087	**-0.565**
Milk price variability	-0.133	0.225	-0.031	0.247	-0.087	-0.014	**0.753**
Cumulative % of the variance explained	24.329	37.776	48.647	57.693	65.344	71.933	**77.419**

Source: calculation based on data of survey in 2010

Extraction method: principal component analysis and loadings larger than 0.5 are in bold

Table 6.5 Varimax rotated factor loading for source of risk

6.4.2 Farmers' risk management strategies

Producing at the lowest cost and preventing or reducing livestock diseases are considered as the most useful risk management strategies in both Inner Mongolia and Hebei Province (Table 6.6).

Category	Total		Hebei Province		Inner Mongolia	
	Mean	Rank	Mean	Rank	Mean	Rank
Producing at the lowest cost	4.76	1	4.81	1	4.68	2
Prevent/reduce livestock diseases	4.76	2	4.76	2	4.75	1
Using consultant service or extension workers	4.6	3	4.64	3	4.54	3
Liquidity - keeping cash at hand	4.24	4	4.17	4	4.35	4
Off-farm work	4.14	5	4.14	5	4.15	5
Collecting information	3.7	6	3.77	7	3.60	6
Joining extension training programme	3.63	7	3.85	6	3.30	8
Shared ownership of equipment, joint operations	3.53	8	3.67	8	3.32	7
Keeping fixed costs low-rent machinery rather than purchase it	3.05	9	3.14	9	2.92	11
Joining farmers' corporation	2.99	10	2.88	10	3.14	9
Production contracts	2.83	11	2.72	11	2.98	10
Asset flexibility - farm building with multiple use	2.68	12	2.72	12	2.63	13
Risk reducing technology	2.63	13	2.45	13	2.88	12
Buying agricultural insurance	2.06	14	2.09	14	2.01	14
Slaughter or sell the cows, quit the business	1.54	15	1.45	15	1.57	15

Source: calculation based on the data of survey in 2010

Note: the order of the risk management is based on the mean score of each one (Column 2)

Table 6.6 Mean score and rank for the risk management strategies

However, some strategies such as production contract, farmer corporation group and agricultural insurance, which are viewed as very important in other countries, were not popular in China, since most dairy farmers in China are still separate and small-scale. In the USA, Patrick and Musser showed that the large-scale US farmers viewed liability insurance as important managerial response to risk [16]. Additionally, the 1996 USDA survey found that keeping cash at hand was the chief risk management strategy for every farm size, for every commodity specialty. This result is also proved in the current research and shows that in China, farmers also consider keeping cash at hand to be a way to deal with all kinds of risk. Quitting the business is listed as the last thing the farmers would do, and it indicates that the farmers would not easily to give up this business when they meet any risk.

As with the sources of risk, the number of risk management strategies was also reduced by applying PCA. The KMO value of risk management strategies is 0.512 and this is acceptable for PCA. Table 6.7 shows the Varimax rotated factor loadings for risk management strategies. This resulted in 6 factors with eigenvalues greater than 1, and total variance explained by the 6 factors accounts for more than 80% of all the variables, which is also considered as satisfactory in social science. Based on the concentration of factor loadings, the six factors can be described as 'cost decrease', 'income stabilization', 'income increase', 'farmer group', 'insurance' and 'consultancy', respectively.

Category	Component					
	Factor 1 Cost decrease	**Factor 2 Income stabilization**	**Factor 3 Income increase**	**Factor 4 Farmer group**	**Factor 5 Insurance**	**Factor 6 Consu-ltancy**
Shared ownership of equipment, joint operations	**0.843**	0.086	0.056	-0.145	0.223	-0.100
Asset flexibility - farm building with multiple use	**0.815**	-0.143	-0.019	-0.008	-0.057	0.116
Risk reducing technology	**-0.719**	0.210	-0.189	0.320	0.108	-0.208
Keeping fixed costs low - rent machinery	**0.582**	0.462	**0.588**	-0.010	0.055	0.080
Production contracts	0.005	**0.893**	0.008	0.028	-0.115	-0.084
Liquidity - keeping cash at hand	-0.211	**0.816**	-0.218	0.326	-0.044	0.127
Off-farm work	-0.024	-0.070	**0.909**	0.074	-0.055	-0.039
Collecting information	0.443	-0.254	**0.641**	0.039	0.156	0.283
Joining the farmers' corporation	-0.069	.058	0.011	**0.937**	-0.004	-0.082
Prevent/reduce livestock diseases	0.119	-0.155	0.211	0.065	**0.814**	-0.195
Buying agricultural insurance	-0.041	0.017	-0.210	-0.106	**0.729**	0.205
Using consultant service or consultant extension workers	0.439	-0.216	-0.089	0.005	-0.009	**0.778**
Joining extension training	0.430	-0.468	-0.216	-0.016	0.092	**-0.665**
Kill or sell the cows, quit the business	0.156	-0.467	-0.189	0.190	-0.071	**-0.684**
Produce at the lowest cost	0.464	-0.156	-0.477	0.096	0.365	0.141
Cumulative % of the variance explained	27.622	43.917	56.236	65.842	73.529	**80.429**

Table 6.7 Varimax rotated factor loadings for risk management strategies

Controlling fixed costs through shared ownership of equipment and partnership loads high on Factor 1 which is 'cost decrease'. However, compared with others, risk reducing technology is outside of the farm business itself, so it shows a negative loading in this factor. Factor 2 and Factor 3 are related to farmers' income. Factor 2, 'income stabilization', has a high loading of product contract which makes a certain income from the farm business itself. Factor 3, 'income increase', has a high loading of off-farm work which keeps a certain income out of the current farm business. Factor 4 is named 'farmer group', only one component is included in this factor. It has a high loading of joining farmers' corporation associations. Factor 5, 'insurance', has high loadings of purchasing agricultural insurance. In addition, preventing or reducing animal disease also shows a high loading on this component. We believe proper treatment and conducting animal health check-up's can reduce the possibility of animal disease, and it can be viewed as a kind of insurance for farmers. Factor 6, 'consultancy', has high loadings of consultant service and extension training in many aspects.

The cluster analysis is based on risk management strategies. Four clusters were assessed by hierarchical cluster analysis using the Ward method. The dendrogram in Figure 6.3 shows the principle and process of hierarchical cluster analysis. All the original cases are sorted into different groups in a way that the distance between two objects is maximal if they belong to the same group and minimal otherwise. Among the methods measuring distances between the objects, the Ward method is regarded as very efficient because it uses an analysis of variance approach to evaluate the distances between clusters. In short, this method attempts to minimize the sum of squares (SS) of any two (hypothetical) clusters that can be formed at each step [17].

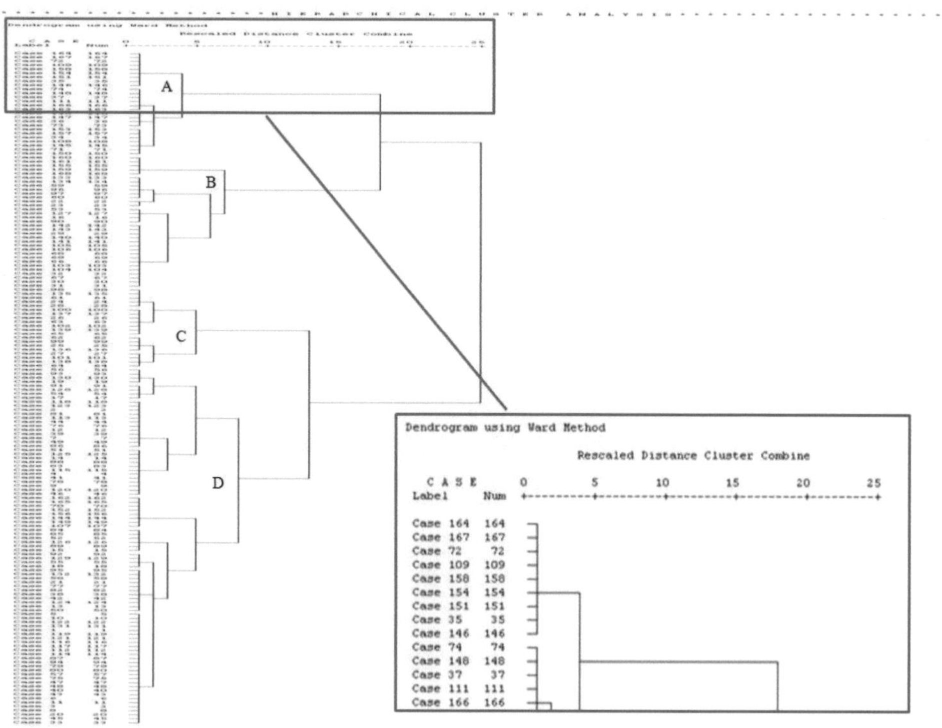

Software: SPSS 13.0

Figure 6.3 Dendrogram of hierarchical cluster analysis

As shown in Figure 6.3, with the increase of distances between the clusters, they are sorted into fewer clusters, until all the cases merge into one cluster with the largest distance of 25. In most of the cases, we need to balance the numbers of clusters and the distances, to analyse the features of the original objects with appropriate scales. Therefore, we decided to choose four clusters and marked them A, B, C and D in Figure 6.3. Table 6.8 shows the different opinions on new risk management strategies among different clusters. As analysed in the last chapter, joining farmers' corporation and production contract are considered new risk management strategies, and others are considered traditional risk management strategies. Farmers in Cluster A are more interested in new strategies such as joining farmers' corporations and using production contracts than farmers in other clusters. Meanwhile, it seems that they are less interested in training programmes compared with farmers in Cluster B, C and D. This implies that farmers in Cluster A are more self-reliant. They rely less on help from training and consultant services, and are more self-reliant.

Cluster	A	B	C	D	All
Shared ownership of equipment, joint operations	2.80	2.93	3.87	3.88	3.52
Join the training programme	2.00	3.96	4.00	3.86	3.63
Joining farmers' corporation	3.96	2.63	2.40	3.13	2.98
Asset-flexibility - farm building with multiple uses	2.00	2.21	2.87	3.05	2.68
Production contracts	3.92	2.36	3.70	2.13	2.82

Table 6.8 Opinion on risk management strategies among different clusters

Table 6.9 shows the comparison among different clusters. Farmers in Cluster A are slightly younger but have a higher education level than other clusters. Meanwhile, farmers in Cluster A usually own more cows than other clusters. All of these facts imply that farmers in Cluster A are younger, better educated and have a larger scale of farm. Farmers with these features are more self-reliant and more interested in new risk management strategies. Put another way, those older, less well educated farmers with smaller scale farms rely more on outside help, training programmes and traditional ways of risk management response.

	Cluster A (26)	Cluster B (33)	Cluster C (40)	Cluster D (69)
Age	41.5	42.9	42.5	41.7
Educated more than 9 years	19.20%	9.20%	10.30%	13.20%
Number of Cows	25	12	14	19

Source: calculation based on the data of survey in 2010

Table 6.9 Comparison of socio-characteristics among four groups

Based on above analysis, Cluster B, Cluster C and Cluster D indicate similar opinions about risk management strategies, and farmers in these three groups show a indicate different ideas than those farmers in Cluster A. Therefore, finally, Cluster B, Cluster C and Cluster D are combined into one large group, called Group 2, Cluster A is called Group 1.

Group	N	%	Features
1	26	15.5	Higher educated, younger with larger scale farms who are more self-reliant, accept new strategies more easily and more willing to join farm cooperation associations
2	142	84.5	Other farmers who rely more on extension and consultant services, and find it hard to accept new strategies

Source: calculation based on the data of survey in 2010

Table 6.10 Different features of the two groups

6.5 Conclusions and policy recommendations

Price and production risks are considered as the most important risk to dairy farmers. Milk price variability is the most important risk to dairy farmers in both Hebei Province and Inner Mongolia. A milk safety issue in the media is a very important risk for dairy farmers in Hebei Province and negatively affects them. Non-domestic epidemic animal diseases, such as foot and mouth disease, are considered as the most serious problems among all kinds of disease.

Based on the above results, we found that the risks to dairy farms can be categorized as 'production risk', 'institutional risk', 'animal disease', 'input market risk', 'milk contamination risk', 'personal risk' and 'output market risk'. Among these risks, price risk and the animal diseases risk are the most serious in dairy farmers' perception. Risk management strategies are categorized as 'cost decrease', 'income stabilization', 'income increase', 'farmer group', 'insurance' and 'consultancy'. Among these strategies, dairy farmers feel that producing at the lowest cost and preventing livestock diseases are the most effective risk management strategies as a single strategy. Dairy farmers in China largely rely on extension or veterinary services, so developing the veterinary extension services to guide farmers to avoid livestock diseases is a good way to reduce these related risks. Extension and veterinary services are good example of 'consultancy'.

Farmer groups and agricultural insurance becomes a way to help farmers to manage risk in crops and some other animal products, however, it is not widely accepted by dairy farmers. Promotion of these strategies might be useful for dairy farmers as well and might help them to avoid many risks and reduce losses.

For the government, more consultant services should be offered; some new risk management strategies should be extended to farmers such as production contracts. For dairy farmers, new risk management strategies, such as agricultural insurance, should be more accepted. The milk powder scandal in China happened in 2008 having a significant negative effect on the local milk industry (mainly in reference to Hebei Province). It totally destroyed the local business and seriously affected public trust in the industry nationwide. With that in mind, even though two years have passed, farmers are still having difficulty selling their milk. A set of effective risk management strategies for dairy farmers is needed.

Although milk price variability is considered as the top risk to dairy farmers, the milk price data is not available in the current research and this is a limitation of the research. In the future, milk market price data should be collected and how milk prices change should be also examined. New risk management strategies are more popular in younger and better educated farmers such as farmers in Group 1 - they usually own more cows and show more self-reliance. These younger and better educated farmers are more interested in joining farm associations, using production contracts with diary enterprises, but are less dependent on extension workers.

Cluster B C and D are similar to each other, so these three clusters are combined into one big group; they rely on training programmes and consultant services more than Group 1. However, those older and less well educated farmers with fewer cattle are not willing to use these new risk strategies. Controlling fixed costs through shared ownership of equipment and partnership loads high on Factor 1 which is 'cost decrease'. However, compared with others, risk reducing technology is outside of the farm business itself, so it shows a negative loading in this factor. Factor 2 and Factor 3 are related to farmers' income. Factor 2, 'income stabilization', has a high loading of production contracts which makes a certain income from the farm business itself. Factor 3, 'income increase', has a high loading of off-farm work which keeps a certain income out of the current farm business. Factor 4 is named as 'farmer group', only one component is included in this factor. It has a high loading of joining farmers' corporation associations. Factor 5, 'insurance', has a high loading of purchasing agricultural insurance. Additionally, preventing or reducing animal diseases also shows a high loading on this component. We believe that by using proper treatment and undertaking animal health checkups can reduce the possibility of animal diseases, and it can be viewed by farmers as a kind of insurance. Factor 6, 'consultancy', has high loadings of consultant service and extension training in many aspects.

References

[1] Nanseki, T. Management of Risk and Information in Agriculture. Agriculture and Forestry Statistics Publishing Inc. Japan (2011). (in Japanese)

[2] FAO (Food and Agriculture Organization)FAOSTAT Agriculture Database, Accessed on November (2008). available at http://faostat.fao.org/site/339/default.aspx

[3] ERS (Economic Research Service)USDA. Agricultural Economic Report Managing Risk in Farming: Concepts, Research (1997). http://www.nal.usda.gov/ref/USDA-pubs/aer.htm(774)

[4] Fleisher, B. (1990). Agricultural Risk Management. Lynne Rienner Publishers Inc. USA.

[5] Wilson P. N., Luginsland T. R., Armstrong D. V. Risk Perceptions and Management Responses of Arizona Dairy Producers. Journal of Dairy Science 1988; 71: 545-551.

[6] Martin S. Risk Management Strategies in New Zealand Agriculture and Horticulture, Review of Market and Agricultural Economics 1996; 64: 31-44.

[7] Meuwissen, M. P. M, Huirne, R. B. M, & Hardaker, J. B. Risk and Risk Management: An Empirical Analysis of Dutch Livestock Farmers. Livestock Production Science 2001; 69: 43-53

[8] Sonkkila, S. Farmers' Decision-Making on Adjustment into the EU: Publication Department of Economics and Management, University of Helsinki, Helsinki (2002). (34)

[9] Wu, Y. Research on the Risk Factors within the Dairy Sector of Inner Mongolia, Thesis of Master Degree, the Inner Mongolia Agriculture University of China (2009). (in Chinese).

[10] Chen, Z. Research on Quality and Safety Control of Dairy Products in China, Thesis of Master Degree, Yangzhou University of China (2009). (in Chinese).

[11] Abdi, H, & Lynne, J. Principal Component Analysis. Interdisciplinary Reviews: Computational Statistics, 2.Chow, W. (2004). An Exploratory Study of the Success Factors for Extranet Adoption in E-supply Chain. Journal of Global Information Management 2010; 12.1: 60-67., 4, 433-459.

[12] Flaten, O., Lien G., Koesling M., Valle P. S., Ebbesvik M. Comparing Risk Perceptions and Risk Management in Organic and Conventional Dairy Farming: Empirical Results from Norway. Livestock Production Science 2005; 95: 11-25.

[13] SPSS IncSPSS Base 15.0 for Windows User's Guide. SPSS Inc., Chicago IL. (2007).

[14] NSBC (Nation Statistical Bureau of China)China Statistical Yearbook (2011). http://www.stats.gov.cn/tjsj/ndsj/2011/indexch.htm.

[15] Chow, W. (2004). An Exploratory Study of the Success Factors for Extranet Adoption in E-supply Chain. Journal of Global Information Management, 12., 1, 60-67.

[16] Patrick G. F., Musser W. N. Sources of and responses to risk: factor analysis of large-scale US corn-belt farmers. In OECD (2000) Income risk management in agriculture. France 1997; OECD: 45-53.

[17] Ward J. H., Jr. Hierarchical grouping to optimize an objective function. Journal of the American Statistical Association 1963; 58 (301): 236-244.

Chapter 7

Consumer Perceptions on Food Safety and Demographic Determinants

Dongpo Li, Tinggui Chen, Hui Zhou and
Teruaki Nanseki

With the fast development of science and technology in food production and processing, food is being supplied to satisfy increasingly diversified tastes, nutritional requirements, etc. Nevertheless, food safety is growing to be a global concern among consumers simultaneously, due to their asymmetric information on the processes, additives in the long industrial chain, and also influence of the flourishing public media. Consumers are demanding the reinforced assurance of food safety and even one isolated event may cause major market disruptions. In addition to the endeavours by governmental agencies and enterprises, food safety, especially from the perspective of promoting consumer confidence, has received the considerable attention of scholars. Thus, in this chapter, the research perspective will be changed from interviewing farmers to consumers, with the further analysis of their perceptions and determining factors towards food safety.

7.1 Introduction

According to the latest research revealed by the international food and grocery expert IGD, China surpassed the US to become the world's largest food and grocery retail market at the end of 2011 [1]. In recent years, especially after the melamine milk powder incident occurred in September 2008 (for more details, refer to Qiao et al. (2012) [2]), many scholars have conducted empirical studies on food safety, based on consumer surveys. In general, the study topics include: (1) consumers' overall perceptions on food safety. Wang et al (2009) [3] and Xu et al (2010) [4] examine consumers' willingness to pay (WTP) for safe fishery products and certified and traceable food, respectively. Qiao et al (2010) [5] studies the changes of consumers' confidence in the domestic dairy industry after melamine milk powder incident, while Zhang et al (2010) [6] divides the sampled consumers into four groups, in respect to their perceptions about and attitudes toward GM food. (2) Consumers' behaviours on choosing safe food. Zhang et al (2010) [7] examines consumers' identification of safe dairy products; Ortega et al (2011) [8] measures the heterogeneity in consumers' preferences for selecting safe pork; Kim (2009) [9] conducts factor analysis on consumers' purchase of GM food. (3) Integrated study of perceptions and behaviours towards food safety. Han et al (2012) [10] compares the consistency of consumers' stated and revealed preferences to certified pork.

Although the existing studies have covered many essential aspects and provided instructive recommendations, there are still a variety of topics that need to be researched further. For instance, (1) consumers' overall awareness of the food safety situation, which constitutes the basis for analysis of individual behaviours; (2) inclusion of questions covering the whole industrial food supply chain, from agricultural production as the origin process; (3) empirical analysis of the relationship between consumers' demographic characteristics and perceptions, etc. Therefore, based on the survey of 512 respondents from Beijing and Shanghai, the top two metropolises in China, this study analyses consumer perceptions towards food safety, including general concern and valuation; major information sources and the subjective reliabilities; awareness about the causes and countermeasures of food safety risks. To explore significant determinants behind the perceptions, a variety of demographic variables with regard to the respondents are included, from gender, age, employment, education background, to the member composition and annual income of each household. The remainder of the chapter is organized as follows: Section 2 briefly describes the questionnaire, sampling and demographic characteristics; Section 3 illustrates the major perceptions of the surveyed consumers; Section 4 analyses the major determinants behind consumer perceptions; Section 5 presents the conclusions and policy recommendations, followed by further discussion.

7.2 Field survey

7.2.1 Questionnaire and sampling

To understand the present situation and farmers' food safety perceptions, we conduct the survey using questionnaire-based personal interviews, to collect first-hand data. The questionnaire consists of 30 multi-choice questions, which are divided into the following four sections, according to different topics of information we intend to collect.

As shown in Table 7.1, (1) we enquire about consumers' overall awareness towards food safety with four questions in the first section. The main topics include their concern about food safety, ranking from *very much* to *not at all*; subjective valuation of the safety of the current food supply, with the main candidate answers ranging from *very safe* to *very risky*; the major source of information, where the candidate answers incorporate not only the traditional mass media of radio and television, newspapers and magazines, and newly developing media of the Internet, but also the other routes such as relatives, experts, commercial advertisements, package information, etc; the information sources deemed to be most reliable vary from government, experts, Internet, relatives, producers and vendors, and commercial advertisement. In addition, with a question for reference, we asked about consumers' awareness of three kinds of certified food: nuisance-free, green and organic. (2) Section 2 is composed of seven questions, within which the first five questions include consumers' awareness on the relationship between environmental pollution and food safety, with 5-level ordinal options from *very intimate* to *does not exist*; most risky substance to food safety, with the multiple choice options of industrial, agricultural and civil pollution; top source of agro-pollution, to be chosen from industrial pollutants, urban or rural civil pollutants, and agricultural chemicals; most risky procedure or stage within the food supply chain, covering agro-production to consumption. In addition, we asked two further questions about consumers' perceptions on the major responsibility bearers of agro-pollution, with the options of the government, farmers, producers of agro-inputting materials (chemicals in the main), consumers, etc.; best ways to control agro-pollution, including the optional answers of legislative perfection, extending environmental technology, etc. (3) Although not adopted in this study, we asked consumers' perceptions on the safety of dairy products with 10 questions in Section 3. The topics vary from overall awareness; subjective valuation on the risky substances and processes; acquaintance of the food certificating systems of GAP (Good Agricultural Practices) and HACCP (Hazard Analysis Critical Control Point); purchasing and consuming behaviours. (4) In the final section, our questionnaire contains eight questions on the demographic characteristics of the respondents. In addition to gender, age, employment and education level, we asked about their family scale, member composition and annual income of the household.

In January to March, 2012, we surveyed consumers in the two metropolises of Beijing and Shanghai. The survey was completed thanks to the kind cooperation of the China Agricultural University and Shanghai Ocean University, from where altogether 30 students were selected and trained as surveyors. All of them are undergraduates or postgraduates majoring in food economics or similar fields. The respondents were determined in two ways: on one hand, they are relatives of the surveyors or people who live near to them; on the other hand, random surveys were conducted through interviewing consumers, encountered mainly near the major supermarkets. In principle, one surveyor can interview no more than 20 consumers. Because some of the authors participated in this survey as well, the initial sample size amounted to 617. Nevertheless, screened mainly by rationality and completion of data, 512 samples are accepted as valid and used in the final analysis, thus the ratio of valid samples is 82.89 %.

7.2.2 Demographic characteristics

1. Basic individual information. In this survey, 304 respondents are female, with the proportion of almost 60%, while the males account for 40.6%. With regard to age distribution, 20-29 year-old respondents have the largest proportion at 30%, followed by the ages of 30-39 (26%), 40-49 (22%), etc.

2. Additional individual information. As to the occupation, respondents employed in enterprises and public institutions account for the largest proportion of 28.7% and 23.2%, respectively. Meanwhile, although categorized as the other types, 21.6% of the sampled consumers are students, retired, etc.; 13.7% are self-employed. In addition, far fewer respondents answered as serving in government or being unemployed. With regard to whether or not they possess a professional background in agriculture, food and medicine, only 14.8% of respondents provided positive answers. Judging from the data collected,

45.9% of the respondents have received a university education and 14.8% are postgraduates; 28.1% having the experience of attending a high school, while respondents with a middle school and less educational background account for only 11.2%.

3. Household information. As to the dichotomous questions on whether a family includes preschool children, primary or middle school students, or people over 60 years, positive answers are 10.2%, 40.8% and 31.1%, respectively. Finally, with respect to the annual household income, one third of the respondents answered as 70-150,000 yuan per year, followed by 35-70,000 yuan per year (28.6%), less than 35,000 yuan per year (17.2%) and 150-300,000 yuan per year (16.4%), while only 4.5 responded as over 300,000 yuan per year (Table 7.1).

	Characteristic	Valid N	%		Characteristic	Valid N	%
1.	**Basic individual information**			2.3	Education level of the respondent (*d*5)	499	100.0
1.1	Gender of the respondent (*d*1)	512	100.0		Primary school and less	10	2.0
	Male	208	40.6		Junior middle school	46	9.2
	Female	304	59.4		High school	140	28.1
1.2	Age of the respondent (*d*2)	504	100.0		Undergraduate	229	45.9
	<20	30	6.0		Postgraduate	74	14.8
	20-29	151	30.0	**3.**	**Household information**		
	30-39	131	26.0	3.1	Preschool Child (*d*6)	512	100.0
	40-49	111	22.0		Yes	52	10.2
	50-59	55	10.9		No	460	89.8
	60 ≤	26	5.1	3.2	Primary and middle school student (*d*7)	512	100.0
2.	**Additional individual information**				Yes	209	40.8
2.1	Employment of the respondent (*d*3)	505	100.0		No	303	59.2
	Government	31	6.1	3.3	Elderly over 60 years (*d*8)	512	100.0
	Public institution [a]	117	23.2		Yes	159	31.1
	Enterprise	145	28.7		No	353	68.9
	Self-employed	69	13.7	3.4	Annual household income (yuan [b] /year, *d*9)	493	100.0
	Jobless	34	6.7		< 35000	85	17.2
	Other (student, retired, etc)	109	21.6		35000-70000	141	28.6
2.2	Background in agriculture, food and medicine (*d*4)	512	100.0		70000-150000	164	33.3
	Yes	76	14.8		150000-300000	81	16.4
	No	436	85.2		300000-500000	18	3.7
					500000 ≤	4	0.8

Note: [a] public institution refers to the institution of public interests, i.e., hospital, educational institutions, academy, etc.; [b] yuan is the major currency unit in China, and 1 US$ equals to 6.30 yuan at the end of 2011.

Source: field survey by the authors

Table 7.1 Demographic characteristics of the surveyed consumers

7.3 Perceptions on food safety

7.3.1 Overall awareness on current situation

1. Concerning food safety. Among 505 valid respondents, altogether 87.5% answer as *very much* or *much* concerned; 11.5% of consumers respond as *a little* concerned; while only 1.0% of consumers *are not* concerned. Similar with the remarks in Xu et al. (2010) [4], Ortega et al. (2011) [8], etc., there is heightened public concern over food safety in China, especially after the outbreak of a series of scandals.

2. Valuation of food safety. Altogether 42.7% of the consumers choose the answer *risky*; the consumers who respond as *safe* amount to 36.8%, followed by 12.9% of consumers answering as *very risky*, while only 3.6% of consumers evaluate the current supply of food as *very safe*. This pessimistic result about food safety is in line with Qiao et al. (2010) [5], where most of the respondents feel unsafe about dairy products.

3. Most used information source. More than half (50.9%) of the interviewed consumers get information concerning food safety from radio and television, followed by the Internet (26.6%), newspapers and magazines (8.7%); each of the other sources chosen as most important information source accounts for no more than 5% among the respondents. The ranking is in accordance with the relative importance of the major media in China, in terms of providing living information to the public in China today. In particular, with the fast popularization of the Internet, it has transcended paper-based newspapers and magazines to be the most important source of information on food safety, for more than one fourth of the sampled consumers.

4. Mostly reliable information source. Although it looks like a similar questions to (3), which focuses on the reality of where consumers get their informed directly, this question intends to capture consumers' subjective judgment of the most reliable information source. According to the answer of 500 valid respondents, government and experts are selected by consumers with the largest proportions of 34.8% and 33.6%, respectively; followed by relatives with a percentage of 10.6%. It is of interests that only 6.2% of respondents chose the Internet, despite 26.6% of interviewed consumers choosing it as the most used information source. Similarly, in the analysis of Zhang et al. (2010) [6], respondents are found to attach high importance to government and scientists, from the perspective of ensuring the safety of genetically modified food (Table 7.2).

7.3.2 On the major affecting factors

1. The relationship of environment and food safety. According to the data collected in this survey, 70.8% of the interviewed consumers believe that there is a *very intimate* relationship between the environment and food safety; 20.3% admit the existence of an *intimate* relationship; while very few respondents deny the influence of the environment on food safety. This indicates that the concept of environmental protection has gained wide acceptance, even from the perspective of food safety.

2. Top threat to food safety. Among the optional answers, industrial pollution is selected by 57.5% of consumers, which is the largest proportion among respondents; followed by agricultural pollution with the proportion of 29.4%; while only 9.1% of the consumers selected civil pollution as the top threat to food safety. In addition to industrial pollution, this result reveals that impacts of agricultural pollution are drawing public attention simultaneously.

3. Top source of agro-pollution. Agricultural chemicals and industrial pollutants are the most significant source of agro-pollution to 45.7% and 39.7% of the interviewed consumers, respectively. Many of the food safety problems in China can be traced back to the farm level, as some farmers still rely heavily on the use of highly toxic pesticides to maintain the output of agro-products [11-12]. Meanwhile, both urban and rural civil pollutants are selected by no more than 8% of the respondents.

4. Most risky procedure or stage of food supply. Among consumers' answer, processing and agricultural production are most risky to food safety, with the proportion of 46.3% and 42.6%, respectively. For food processing, the high ratio may due to asymmetric information on the operations, additives and the frequent disclosure of related scandals by the public media [2, 8]. Simultaneously, far fewer respondents selected the other procedures or stages, i.e., harvest of agro-products (2.4%), transportation (4.8%), marketing (1.4%), consumption (0.2%), etc. (Table 7.2).

7.3.3 On the risk management

1. Major responsibility bearer of agro-pollution. Most of the surveyed consumers ascribe the responsibility to government, followed by producers of agro-chemicals, etc., with the proportions at 54.0% and 33.6%, respectively. In previous literature, Dellios et al (2009) [13] introduces the concept of corporate social responsibility (CSR) into the food industry, and explores the important role of government in tackling food safety problems. Similarly, Qiao et al (2010) [5] analyses the responsibilities of government and corporate agencies, in taking efficient countermeasures to ensure food safety and maintain consumer confidence. In addition, far fewer respondents believe that this responsibility should be borne by farmers (5.7%) and consumers (1.4%), etc.

2. Best ways to control agro-pollution. The answers to which can be divided into three groups. The first group includes perfecting the legislation, extending environmental technology, supported by 33.2 and 26.9 % of respondents, respectively. As surveyed by Li et al. (2012) [2], quite a few (12.4%) consumers have knowledge about the Act of Food Safety that came into effect on June 1, 2009. According to our prior survey, toxic pesticides are still being widely used, while bio-controls of pests are not yet well extended to agricultural production [12]. The second group comprises strengthening the penalties, enlarging farm managerial scales, with the largest proportion of 19.0% and 10.6%, respectively. The significance of the larger farming scales lies in the easier adoption of advanced and environmental technology, through initiatives and capital power. The third group consists of subsiding environmental behaviours and others, as viewed by 7.3% and 3.0% of surveyed consumers, respectively (Table 7.2).

	Characteristic	N	%		Characteristic	N	%
1.	**Overall awareness**			2.2	Top threat to food safety	504	100.0
1.1	Concern about food safety	505	100.0		Industrial pollution	290	57.5
	Very much	218	43.1		Agricultural pollution	148	29.4
	Much	224	44.4		Civil pollution	46	9.1
	A little	58	11.5		No idea	20	4.0
	Not at all	5	1.0	2.3	Top source of agro-pollution	501	100.0
1.2	Valuation of food safety	505	100.0		Industrial pollutant	199	39.7
	Very safe	18	3.6		Urban civil pollutant	37	7.4
	Safe	186	36.8		Rural civil pollutant	34	6.8
	Risky	216	42.7		Agricultural chemicals	229	45.7
	Very risky	65	12.9		Other	2	0.4
	No idea	20	4.0	2.4	Most risky stage	505	100.0
1.3	Most used information source	493	100.0		Agricultural production	215	42.6
	Radio & television	251	50.9		Harvest of agro-products	12	2.4
	Newspaper & magazine	43	8.7		Processing	234	46.3
	Internet	131	26.6		Transportation	24	4.8
	Relatives	24	4.9		Marketing	7	1.4
	Experts (doctor, researcher, etc.)	12	2.4		Consumption	1	0.2
	Commercial advertisement	7	1.4		Other	11	2.2
	Package information	21	4.3		No idea	1	0.2
	Other	4	0.8	**3.**	**On the risk management**		
1.4	Most reliable information source	500	100.0	3.1	Major responsibility bearer of agro-pollution	506	100.0
	Government	174	34.8		Government	273	54.0
	Experts (doctor, researcher, etc.)	168	33.6		Farmer	29	5.7
	Internet	31	6.2		Producer of agro-chemicals	170	33.6
	Relatives	53	10.6		Consumer	7	1.4
	Producers & vendors	11	2.2		No idea	15	3.0
	Commercial advertisement	29	5.8		Other	12	2.3
	None	29	5.8	3.2	Best way to control agro-pollution	506	100.0
	Other	5	1.0		Perfecting the legislation	168	33.2
2.	**On the major affecting factors**				Technological Extension	136	26.9
2.1	Environment and food safety	503	100.0		Subsiding good behaviours	37	7.3
	Very intimate	356	70.8		Strengthening the penalties	96	19.0
	Intimate	102	20.3		Enlarging farming scales	54	10.6
	No idea	40	8.0		Other	15	3.0
	Almost no	3	0.6				
	Not exist	2	0.4				

Source: field survey by the authors

Table 7.2 Summary of consumer perceptions on food safety

7.4 Impact of demographic characteristics

7.4.1 Significance of demographic effects

To measure the relative importance of the candidate perceptions, we calculate their ratios within each demographic variable. Furthermore, similar to Steiner et al. (2005) [14] and Gacula et al. (2006) [15], coefficient of variation (CV) of these ratios is computed to showcase a discrepancy of consumer perceptions towards a certain optional answer about food safety, crossing different features within each demographic characteristic (Table 7.3). Taking the first value of 0.17 as an instance, this is the CV of percentages of male and female respondents concerning food safety *very much*. In general, a smaller value of CV indicates less variation of responding ratios, hence less influence from the difference of this demographic characteristic. Another instance relates to the ratios of concern about food safety *very much*, where the CVs from different gender and age are 0.17 and 0.14, respectively, thus the affect of gender is larger than that of the age. In succession, to identify the significance of the CVs, a one-way T-test is conducted with the application of SPSS 13.0 [14][1]. The null hypothesis is that each population mean of CVs is not significantly different from 0, when respondents' choices are unaffected by this demographic characteristic. If the null hypothesis is rejected with a smaller p-value than the thresholds of 0.01, 0.05 or 0.1, then there is evidence that a significant discrepancy exists among different features within a demographic characteristic, and vice versa [16].

Perception and options [a]		*d1*	*d2*	*d3*	*d4*	*d5*	*d6*	*d7*	*d8*	*d9*
Concern about food safety	Very much	0.17	0.14	0.17	0.08	0.26	0.02	0.03	0.07	0.32
	Much	0.09	0.12	0.07	0.10	0.18	0.03	0.04	0.03	0.21
	A little	0.20	0.39	0.52	0.07	0.59	0.00	0.15	0.13	0.54
	Not at all	0.99	1.22	1.08	0.82	0.96	1.41	1.41	0.41	1.11
...[b]	...	...	...	...	...	...	...	...	...	...
Best way to control agro-pollution	Perfecting the legislation	0.12	0.13	0.23	0.19	0.39	0.16	0.13	0.03	0.49
	Extending environmental technology	0.02	0.39	0.23	0.23	0.09	0.00	0.16	0.05	0.27
	Subsiding environmental behaviours	0.06	0.61	0.55	0.06	0.56	0.04	0.02	0.26	0.71
	Strengthening the penalties	0.17	0.42	0.56	0.09	0.32	0.22	0.21	0.15	0.55
	Enlarging farm managerial scales	0.07	0.46	0.28	0.12	0.36	0.08	0.34	0.13	0.63
	Other	0.24	1.31	1.31	0.70	1.13	0.42	0.24	0.51	2.06

Note: [a] responses of *no idea* are excluded in this table; [b] the same coefficient is computed within other perceptions

Software: SPSS 13.0

Table 7.3 Coefficient of variation of responding ratios within each characteristic

1 The R2 value of 0.406 should not be used to judge fitness of a model. The fact that R2 never decreases when any variable is added to a regression makes it a poor tool for deciding whether one or several variables should be added to a model. Low R2s in regression equations are not uncommon, especially for cross-sectional analysis. Thus using R2 as the main gauge of success for an econometric analysis can lead to difficulties [17].

	d1	d2	d3	d4	d5	d6	d7	d8	d9
Concern about food safety	1.72	1.81	2.02	1.45	2.80*	1.05*	1.22	1.86	2.72*
Valuation of food safety	2.67*	2.18	1.92	2.55*	3.44**	2.06	1.30	1.54	4.17**
Most used information source	4.14***	4.75***	3.33**	2.76**	5.66***	2.84**	3.35**	2.28*	5.45***
Most reliable information source	3.85***	4.94***	4.14***	2.89**	5.47***	1.76	3.29**	2.78**	6.71***
Relationship of environment and food safety	3.84**	2.51*	2.36*	2.17*	3.30**	2.14*	2.29*	1.94	2.18*
Top threat to food safety	2.50	3.37*	2.18	3.39*	5.63**	1.39	1.49	8.49**	4.39**
Top source of agro-pollution	1.74	2.38**	1.52	2.63*	2.85**	1.87	1.34	1.29	2.51*
Most risky procedure or stage of food supply	2.51**	2.62**	3.26**	2.78**	3.68**	2.98**	1.87	2.72**	3.42**
Major responsibility bearer of agro-pollution	2.62*	2.93**	2.86**	1.63	3.22**	1.48	2.70*	1.96	5.40***
Best way to control agro-pollution	3.43**	3.37**	3.12**	2.38*	3.28**	2.45*	4.15***	2.60**	2.99**

Note: [a] numerals are the T-values of the one-sample T-test on whether the mean of CVs within each perception significantly differs from 0; [b] ***, ** and * denote statistical significance in the level of 0.01, 0.05 and 0.1 respectively

Software: SPSS 13.0

Table 7.4 T-values of the discrepancy of response with regard to different characteristics

As shown in Table 7.4, education level of the respondent (*d*5) and annual household income (*d*9) are significant, despite the differences in significance level, in capturing respondents' discrepancy on options towards all the 10 types of perceptions. Meanwhile, all the other seven demographic variables are measured as significant in identifying discrepancies among most of the perceptions. Thus, in our questionnaire, the adoption and classification of the demographic variables are rational, in showcasing the diversified perceptions among surveyed consumers.

	Gender (d1)		Age of the respondent (d2)					
	Female	Male	<20	20-29	30-39	40-49	50-59	60≤
Radio & television	50.0	52.2	26.7	45.8	45.7	56.8	66.0	80.8
Newspaper & magazine	8.3	9.4	3.3	8.5	10.9	12.6	3.8	0.0
Internet	27.9	24.6	60.0	33.1	29.5	15.3	15.1	7.7
Relatives	5.2	4.4	0.0	3.5	6.2	3.6	9.4	7.7
Experts	2.1	3.0	6.7	2.1	2.3	2.7	1.9	0.0
Commercial advertisement	1.4	1.5	0.0	0.7	1.6	3.6	0.0	0.0
Package information	4.5	3.9	3.3	4.9	3.1	5.4	3.8	3.8
Other	0.7	1.0	0.0	1.4	0.8	0.0	0.0	0.0
#Valid N	290	203	30	142	129	111	53	26

	Education level of the respondent (d5)					Annual household income (1000 yuan) (d9)					
	Primary and less	Junior middle	High school	Undergraduate	Post-graduate	<35	35-70	70-150	150-300	300-500	500≤
Radio & television	60.0	67.4	58.3	47.5	38.4	63.9	52.5	47.8	44.3	50.0	25.0
Newspaper & magazine	0.0	4.3	8.6	10.5	8.2	6.0	10.1	11.5	6.3	5.6	0.0
Internet	10.0	2.2	18.0	33.3	37.0	19.3	23.7	26.1	34.2	33.3	75.0
Relatives	20.0	8.7	5.0	2.7	6.8	3.6	4.3	5.7	6.3	5.6	0.0
Experts	0.0	2.2	1.4	2.3	5.5	0.0	2.2	3.2	2.5	5.6	0.0
Commercial advertisement	0.0	2.2	3.6	0.5	0.0	1.2	2.2	1.9	0.0	0.0	0.0
Package information	10.0	10.9	5.0	2.3	4.1	4.8	4.3	3.8	6.3	0.0	0.0
Other	0.0	2.2	0.0	0.9	0.0	1.2	0.7	0.0	0.0	0.0	0.0
#Valid N	10	46	139	219	73	83	139	157	79	18	4

Table 7.5 Percentages of responses to the major sources of information

7.4.2 Effect on the overall awareness

Based on the results of the T-test in Table 7.4, it makes sense to conduct further analysis on consumers' perception, taking into account the significant effects of demographic characteristics. Nevertheless, to capture the major determinants of consumers' perceptions, only relationships at the level of 0.01 are included below.

1. Most used information source is significantly determined by the following variables. i) Gender of the respondents (*Gender*), where males get more information from the traditional media of radio, television, newspapers and magazines, while females are relying more on information from the Internet and relatives. ii) Age of the respondent (*d*2) is found

as being positively correlated with the use of radio and television, and negatively correlated with the Internet, respectively. iii) For education level of the respondent (*d*5), negative relationships are found with the use of radio and television, while there is a positive correlation with the Internet. iv) Annual household income (*d*9) is measured as being negatively correlated with radio and television, while positively correlated with the Internet, similarly (Table 7.5). These findings are in line with the reality that the Internet is usually used for gathering information by young people with more leisure time, who demand a fashionable lifestyle and who tend to be better educated. However, the traditional media are still important in affecting consumers' perception and behaviour.

2. Most reliable information source. i) On the significant variable of *Gender*, 40.4% of male respondents answer as being convinced by information released by the government in the first place, while 37.3% of the females believe experts to be the most reliable source of information concerning food safety. ii) In terms of the *Age* variable, consumers aged 50-59 are sampled as having greatest faith in government and relatives, while those aged 40-49 choose the experts. In addition, although proportioned only 10.1%, the 20-29 aged consumers responded as having most faith in the Internet. iii) Analysing from the perspective of employment, self-employed consumers and those working for the government are identified as possessing the top proportions of those who believe in the information issued by the government and experts, respectively. iv) With regard to the impacts of education, consumers with middle-level educational background show most reliance on government, while consumers who have received higher levels of education have most confidence in the information given by experts. v) As to the effects of income, a negative relationship is found with the percentage of reliance on government. Meanwhile, households with the total annual income over 500,000 yuan have the largest proportion of trust in experts in related fields (Table 7.6).

7.4.3 Effect on the perceptions of risk management

1. Major responsibility bearer of agro-pollution. According to the result of the T-test, the largest discrepancies are identified across the annual household *income* of the consumers. In detail, those with the annual income over 300,000 yuan have the largest proportion of at least 70.6%, attributing the major bearer to government (Table 7.7). This may because people with much more income than average have the high-level demands of life, with food safety included as the basic requirement. As the Chinese government has substantial power on almost all the sectors of the national economy, it is indispensible to rely on the function of government to ensure food safety.

2. Best ways to control agro-pollution. Among the demographic characteristics, only whether there is primary or middle school student in a household (*d*7) is measured as significant at the level of 0.01. The major reasons behind this may include that in the two metropolises, where almost all families have only one child, who is treated preciously by the whole family. Moreover, children enrolled at primary or middle school are at a critical phase of physical and mental development. Different from preschool children who intake food supplied mainly by the family or kindergarten, primary and middle school students

often buy food outside and intake it by themselves. Thus, their food safety is drawing significant attention from both the family and society. For the respondents with students in their family, the largest proportion of 37.6% supported the perfecting of related legislation, which is larger than that of negative respondents. Meanwhile, more consumers without students in the family answered in favour of extending environment-friendly technologies (29.8%) and strengthening the penalty on behaviours undermining the agricultural environment (21.6%), being 5.9% and 4.5% higher than that of the consumers with a student family member, respectively (Table 7.8).

	Gender (d1)		**Age of the respondent (d2)**						**Employment of the respondent (d3)**					
	F.	**M.**	**<20**	**20 -29**	**30 -39**	**40 -49**	**50 -59**	**60≤**	**Gov.**	**Public ins.**	**Enter- prise**	**Self employed**	**Job- less**	**Other**
Government	30.8	40.4	37.9	25.5	37.2	40.5	40.7	38.5	35.5	34.2	33.6	42.6	35.3	32.1
Experts	37.7	27.9	34.5	33.6	34.1	37.8	24.1	30.8	38.7	41.0	33.6	27.9	26.5	29.2
Internet	6.8	5.3	3.4	10.1	5.4	0.9	9.3	7.7	0.0	3.4	4.9	10.3	2.9	11.3
Relatives	10.3	11.1	10.3	10.1	9.3	9.0	18.5	11.5	16.1	7.7	9.1	10.3	17.6	12.3
Producers & vendors	2.4	1.9	3.4	4.0	1.6	0.9	0.0	3.8	0.0	0.9	1.4	4.4	2.9	3.8
Commercial advertisement	5.1	6.7	0.0	6.0	5.4	8.1	3.7	7.7	6.5	6.8	6.3	2.9	11.8	3.8
None	5.8	5.8	10.3	9.4	6.2	1.8	1.9	0.0	3.2	5.1	9.1	1.5	2.9	6.6
Other	1.0	1.0	0.0	1.3	0.8	0.9	1.9	0.0	0.0	0.9	2.1	0.0	0.0	0.9
#Valid N	292	208	29	149	129	111	54	26	31	117	143	68	34	106

	Education level of the respondent (d5)					**Annual household income (1000 yuan) (d9)**					
	Primary and less	**Junior middle**	**High school**	**Undergraduate**	**Post-graduate**	**<35**	**35 -70**	**70 -150**	**150 -300**	**300 -500**	**500≤**
Government	30.0	55.6	42.0	30.5	25.7	41.7	37.1	33.8	32.1	22.2	0.0
Experts	20.0	20.0	29.0	38.1	36.5	21.4	40.0	34.4	35.8	22.2	50.0
Internet	20.0	4.4	2.9	6.6	9.5	6.0	7.1	5.6	3.7	22.2	0.0
Relatives	20.0	11.1	14.5	7.5	10.8	13.1	6.4	10.6	9.9	27.8	0.0
Producers & vendors	10.0	4.4	1.4	1.8	2.7	6.0	1.4	0.6	3.7	0.0	0.0
Commercial advertisement	0.0	4.4	5.1	5.8	9.5	4.8	4.3	6.9	7.4	5.6	25.0
None	0.0	0.0	3.6	8.4	5.4	7.1	2.1	6.3	7.4	0.0	25.0
Other	0.0	0.0	1.4	1.3	0.0	0.0	1.4	1.9	0.0	0.0	0.0
#Valid N	10	45	138	226	74	84	140	160	81	18	4

Table 7.6 Percentages of responses to the most reliable sources of information

	Annual household income (1000 yuan) (d9)					
	<35	35-70	70-150	150-300	300-500	500≤
Government	59.3	51.1	55.6	53.2	70.6	100.0
Farmer	3.7	6.6	8.0	5.2	0.0	0.0
Producer of agro-chemicals, etc.	30.9	38.0	34.0	39.0	23.5	0.0
Consumer	2.5	1.5	0.6	1.3	0.0	0.0
Other	3.7	2.9	1.9	1.3	5.9	0.0
# Valid N	81	137	162	77	17	4

Source: field survey by the authors

Table 7.7 Percentages of responses to the major responsibility bearer of agro-pollution

	Primary and middle school students (d7)	
	Yes	No
Perfecting the legislation	31.2	37.6
Extending environmental technology	29.8	23.9
Subsiding environmental behaviours	7.5	7.3
Strengthening the penalties	21.6	16.1
Enlarging farm managerial scales	8.6	14.1
Other	1.4	1.0
#Valid N	292	205

Source: field survey by the authors

Table 7.8 Percentages of responses to the best ways to control agro-pollution

7.5 Conclusions and recommendations

7.5.1 Major conclusions

Based on the survey of 512 respondents from Beijing and Shanghai, this chapter studies consumer perceptions on food safety. (1) Analyses on the overall awareness indicate that most of the interviewed consumers are concerned about food safety; more than half of the respondents think the current situation of food supply is risky or very risky; the television and Internet are the most important sources of information on food safety, while information from the government and experts are deemed most reliable. (2) In terms of perceptions on the major affecting factors, the significance of environmental protection in ensuring food safety has been accepted by more than 90% of the surveyed consumers. Industrial and agricultural pollution are thought to be the top threat to food safety by almost 90% of respondents, while agro-chemicals and industrial pollutant are the top source of agro-pollution, at more than 80% of the sampled consumers. Processing and agricultural production are the most risky procedure or stage, with the proportion of 46.3% and 42.6%, respectively. (3) As to the risk management of food safety, most of the respondents think that the government and producers of the materials inputted to agriculture, especially those of agro-chemicals, should take the responsibilities for agro-pollution at first. While 60% of the consumers believe perfecting the legislation and extending environmental technology are the best ways to control agro-pollution.

The T-test reveals that all the nine demographic variables are significant in identifying discrepancies among most of the perceptions. (1) With respect to the most important information source, the Internet is more used for gathering information by young people with more leisure time, demands for a fashionable lifestyle and who tend to be better educated. Meanwhile, the traditional media of television, newspapers and magazines are still important in affecting consumer perception and behaviour. (2) From the perspective of most reliable information source, the government is supported more by male, older and self-employed consumers, while experts are trusted by more female, middle-aged, public servants and rich consumers. (3) The government is attributed taking the responsibility for agro-pollution significantly by those with an annual income over 300,000 yuan. (4) Whether there is primary or middle school student in a household is measured as a high-significantly factor affecting respondents' attitude toward the best ways to control agro-pollution.

7.5.2 Policy recommendations

(1) Since the government is chosen as the most reliable source of information and the major responsibility bearer of agro-pollution by most respondents, its supervisory obligations on food safety should be strengthened. Under the unified coordination and leadership of the National Food Safety Committee[2], responsibility for each department needs to be clarified, hence improving the supervisory efficiency through more initiatives and regular investigations. (2) Considering the fact that most of the consumers responded as concerned and worried about food safety, the results of the supervisions should be disclosed to the public promptly. (3) Based on the findings of most risky procedure or stage, key sectors of food safety supervision include: appropriate sterilization, additives and labelling in the processing operations of food manufacturers, proper manufacturing and application of agro-chemicals. (4) As to the Internet which is found to be widely used but no so trusted, further inspections are needed to improve its reliability in releasing food safety information. On the other hand, the government should make full use of mass media in collecting information and communicating with the public. (5) Being one of the best ways to control agro-pollution, relevant legislations are necessary, including further provisions in the existing acts and formulation of an Act of Agricultural Pollution Prevention, etc. (6) Due to the intimate relation between the environment and agro-pollution, and the functions in controlling agro-pollution, extension of environment-friendly techniques needs to be accelerated.

2 According to provision on the Act of Food Safety (2009), the National Food Safety Committee was established in the State Council on February 9 2010. As the top administrative agency of food safety, this committee is headed by the outstanding vice-premier and other two vice-premiers, and composed of 15 concerned ministries.

7.5.3 Further discussion

Food safety will continue to be a public concern in China. In addition to the increased demand for healthy and safe foods with the rapidly growing economy, it can be attributed to the frequent occurrence of scandals. As we write this manuscript April 2012, large volumes of jellies and medicine capsules have been found to contain excessive amounts of chromium. In this wide-ranging incident, some famous companies are even involved, and the major substance concerned is suspected of being illegally added industrial gelatin. For the government and domestic food industry, they still have to strive for food safety. Meanwhile, a variety of topics are open to academic research, in terms of exploring technologies and countermeasures to ensure food safety. For instance, comparative study on behaviours and perceptions between producer and consumers, on different foods or cultures, are beneficial for the administration of food safety.

References

[1] Askew K. Global: China becomes world's largest food market. Just food. 2012: http://www.just-food.com/news/china-becomes-worlds-largest-food-market_id118784.aspx.

[2] Qiao G., Guo T., Klein K. K. Melamine and other food safety and health scares in China: Comparing households with and without young children. Food Control, 2012; 26: 378-386.

[3] Wang F., Zhang J., Mu W., Fu Z., Zhang X. Consumers' perception toward quality and safety of fishery products, Beijing, China. Food Control, 2009; 20: 918-922.

[4] Xu L., Wu L. Food safety and consumer willingness to pay for certified traceable food in China. J. Sci. Food & Agri. 2010; 90: 1368-1373.

[5] Qiao G., Guo T., Klein K. K. Melamine in Chinese milk products and consumer confidence. Appetite, 2010; 55: 190-195.

[6] Zhang X., Huang J., Qiu H., Huang Z. A consumer segmentation study with regards to genetically modified food in urban China. Food Policy, 2010; 35: 456-462

[7] Zhang C., Bai J., Lohmar B. T., Huang J. How do consumers determine the safety of milk in Beijing, China? China Econ. Rev. 2010; 21: S45-S54.

[8] Ortega D. L., Wang H. H., Wu L., Olynk N. J. Modeling heterogeneity in consumer preferences for select food safety attributes in China. Food Policy 2011; 36: 318-324.

[9] Kim R. B. Factors influencing Chinese consumer behavior when buying innovative food products. Agr. Econ. - Czech 2009; 55: 436-445.

[10] Han Q., Zhou H., Nanseki T., Wang J. The consistency of consumer's stated preference and revealed preference: evidence from agricultural product market in China, J. Fac. Agr., Kyushu Univ. 2012; 57 (1): 227-234.

[11] Zhou J., Jin S. Safety of vegetables and the use of pesticides by farmers in China: Evidence from Zhejiang Province. Food Control 2008; 20: 1043-1048.

[12] Li D., Nanseki T., Takeuchi S., Song M., Chen T., Zhou H. Farmers' behaviors, perceptions and determinants of pesticides application in China: evidence from six eastern provincial-level regions. J. Fac. Agr., Kyushu Univ. 2012; 57 (1): 255-263

[13] Dellios R., Yang X., Yilmaz N. K. Food safety and the role of the government: implications for CSR policies in China. i Business 2009; 1: 75-84

[14] Steiner C. F., Darcy-Hall T. L., Dorn N. J., Garcia E. A., Mittelbach G. G, Wojdak J. M. The influence of consumer diversity and indirect facilitation on trophic level biomass and stability. OIKOS 2005; 110: 556-566

[15] Gacula M. Jr., Rutenbeck S. Sample size in consumer test and descriptive analysis. J. Sens. Stud. 2006; 21 (2): 129-145

[16] Bruin J. Newtest: Command To Compute New Test. UCLA: Academic Technology Services. Statistical Consulting Group 2006: http://www.ats.ucla.edu/stat/spss/output/Spss_ttest.htm

[17] Wooldridge J. M. Introductory Econometrics: A Modern Approach (2nd Edition). South-Western Thomson Learning, Mason 2003: 41, 81.

Chapter 8

Consumers' Risk Awareness and Willingness to Pay for Certified Food

Hui Zhou and Teruaki Nanseki

8.1 Introduction

8.1.1 Development of the dairy industry

The dairy industry has huge potential in China. The production and consumption of milk in China have increased dramatically, especially since 2000. However, recent food safety problems which have occurred in the livestock sector in China have negatively affected consumers' confidence in purchasing foods. Food safety problems in the dairy industry have created lack of confidence among the public in buying dairy products.

New approaches to ensure food safety, such as traceability system, good agricultural practices (GAP), hazard analysis and critical control point (HACCP), are of concern to the government, producers and even consumers. In Japan, Aizaki has evaluated Japanese consumers' willingness to pay for GAP certificated milk [1]. The traceability system could be a new way to respect consumers' right to choose safe and quality food, and to provide production information to consumers which may help improve consumers' confidence in the food they consume. Basically, one of the functions of a traceability system is to monitor food producers in order to avoid food safety problems such as misuse of veterinary medicines and so on. Therefore, in this chapter, the traceability system is viewed as a main example, to examine consumers' awareness of food risk and willingness to pay for food safety certification.

At present, the traceability system can be found in some supermarkets and hypermarkets in China and it is mainly used on vegetables (Figure 8.1). The traceability system is a new thing to Chinese consumers and consumer attitudes toward the traceability system are not as yet clear. In this chapter, the main research objectives are to examine consumers' purchasing behaviour on milk, to study consumers' response to food safety issues and to investigate the consumers' awareness of the traceability system.

Source: photographs by the authors

Figure 8.1 Food traceability system in Beijing

8.1.2 Food safety certification

Food safety and food quality have increasingly come to the forefront of consumer concerns, industry strategies and government policy initiatives. In recent years, a number of serious food safety problems within China have negatively affected consumer confidence on both domestic and exported food products.

New approaches to ensure food safety are becoming more integrated, in order to uphold consumers' rights to know about the commodities they consume. A traceability system can be one efficient way to advocate the prompt and accurate movement of information covering all processes in the food supply, so that the public have confidence to choose and consume food. Basically, a traceability system is a tool to monitor food producers in order to avoid food safety problems such as illegal additives, misuse of veterinary drugs, occurrence of animal diseases, etc.

The traceability system is still new to Chinese consumers, whose attitudes towards the traceability system have been examined. In this chapter, the production history in the traceability system is mainly studied, due to the findings that most consumers are concerned about the production process in both Japan and China [2]. The main objectives are to study consumers' marginal willingness to pay (MWTP) on the information that the traceability system provides, and to determine factors affecting consumers' willingness to pay for a traceability system.

8.2 Literature review

As a hot topic, traceability system has been studied for several years. Several research groups have studied the global requirements and impacts of a traceability system [3, 4]. Meanwhile, some other authors, such as Sahin E., et al [5], considered information technology (IT) to be the fundamental tool to assist bringing about revolutionary changes in product traceability.

There is much research mainly focusing on consumer behaviour-consumer opinions on the traceability system applied to the meat supply chain in Europe, America and Canada [6]. US consumers were willing to pay $0.23 more on beef hamburgers with traceability assurance, $0.50 to add assurances on animal treatment, $0.63 to add extra assurances of food safety, and $1.06 to upgrade the sandwich to one which contains all three upgrades. For pork, the same respective upgrades were valued on the average at $0.50, $0.53, $0.59 and $1.1, respectively [6]. The consumers in Canada were willing to pay non-trivial amounts (around 0.33$) for beef with a traceability assurance. However, the quality assurance with respect to food safety and on-farm production methods for beef were more valuable to consumers than simple traceability system assurance. Bundling traceability with both of these quality assurances yielded a $0.83 premium over the traceability-only beef [7].

However, in China, there has only been limited research in terms of economics. After Bird Flu occurred in China, the consumers in Beijing were willing to pay 0.80-1.53 yuan/kg more for chicken with a traceability label if they had more knowledge about the traceability label while the price of chicken was 9.40 yuan/kg [8].

8.3 Data collection and research method

8.3.1 Research areas

To examine consumers' attitude toward the traceability system, an interview survey was conducted from September to October 2008 in Beijing, with 209 valid samples collected. Meanwhile, data from another field survey in Beijing, July 2008, conducted by Nanseki et al. (2008) [9] is also used for analysis in this chapter. In this survey, 214 consumers were interviewed and their answers collected as valid samples.

The total population of Beijing amounted to 17.15 million by 2010, within which some 11.9 million are registered permanent residents, and around five million are temporary residents. Beijing consists of 16 districts (Figure 8.2), and based on the location and socio-economic features, they are categorized into four regions (Table 8.1), according to the capital planning of main functional regions released in July, 2012 [10]. As the Capital and one of the biggest cities in China, Beijing has one of the highest averages of milk consumption zones in China (Figure 8.3). In particular, thanks to the Olympic Games held in the summer of 2008, the traceability system has been applied in some supermarkets in Beijing. Therefore, we chose Beijing as the research area in this chapter.

Source: revision based on http://www.chinahighlights.com/beijing/map.htm

Figure 8.2 Location and districts of Beijing

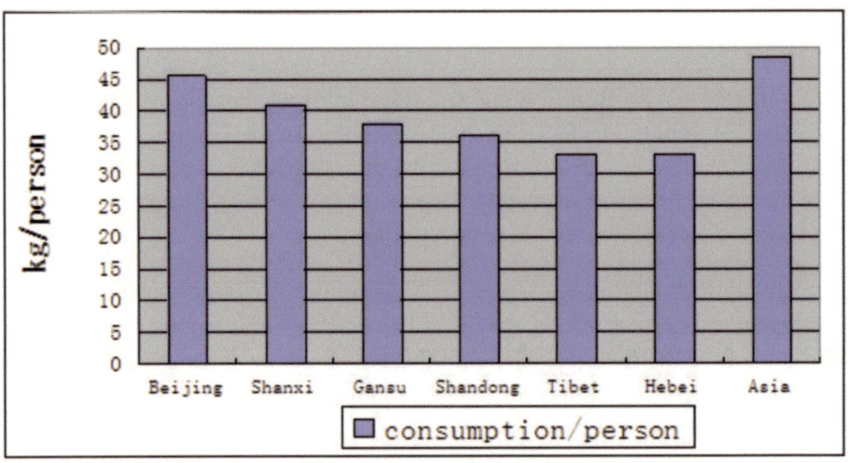

Source: China Ministry of Agriculture (2012) [11]

Figure 8.3 Consumption of dairy products in some regions

Districts	Registered permanent residents			Temporary residents
	Total	Non-Agricultural	Agricultural	
Total	**1197.6**	**905.4**	**292.2**	**516.9**
Core districts of capital function	**225.3**	**225.3**		**40.4**
Dongcheng	95.7	95.7		19.6
Xicheng	129.6	129.6		20.8
Urban function extended districts	**508.2**	**469.7**	**38.5**	**318.9**
Chaoyang	174.5	159.6	14.9	130.6
Fengtai	99.6	86.6	13.0	80.4
Shijingshan	35.2	35.2		15.4
Haidian	198.9	188.3	10.6	92.5
New districts of urban development	**302.4**	**138.9**	**163.5**	**133.2**
Fangshan	75.8	36.4	39.4	16.1
Tongzhou	63.7	28.6	35.1	24.2
Shunyi	56.2	22.3	33.9	21.9
Changping	49.2	26.9	22.3	35.5
Daxing	57.5	24.7	32.8	35.5
Ecological preservation districts	**161.7**	**71.5**	**90.2**	**24.4**
Mentougou	23.9	17.4	6.5	8.1
Huairou	27.4	11.1	16.3	7.0
Pinggu	39.7	16.9	22.8	2.6
Miyun	42.9	15.8	27.1	4.5
Yanqing	27.8	10.3	17.5	2.2

Sources: r Official Website of the Beijing Government (2012) [10]

Table 8.1 Population distribution in Beijing (2010) (Unit: 10000 persons)

8.3.2 Survey and data

This is a chapter, in order to research consumers' attitudes towards milk purchasing, and milk safety and quality, an interview survey was conducted in September and October 2008. The study areas cover seven urban districts, which share 83% of the population of Beijing. According to the population allocation of each district, we choose a certain number respondents randomly from each district. The survey was carried out in the four main supermarkets or hypermarkets. According to the population distribution, also because of limited of transportation, seven districts which include 60% of the population were chosen to carry out the survey (Table 8.2).

Location	N	%
Haidian	60	28.7
Chaoyang	53	25.3
Dongcheng	18	8.5
Xicheng	22	10.6
Xuanwu *	17	8.2
Chongwen *	10	4.8
Fengtai	29	13.8
Total	209	100.0

Note: after the district adjustment of 2010, Xuanwu and Chongwen were merged into Xicheng and Dongcheng Districts, respectively.

Source: consumer survey in September, 2008

Table 8.2 Data distribution in each district

To understand the factors affecting consumers' purchase of milk and other dairy products, it is significant to interview not only the dairy producers, but also the policy-makers.

	Category	%		Category	%
Gender	Male	31.2	Age	Under 18	1.4
	Female	68.8		19-25	23.4
Household income	<1000	3.8		26-35	29.1
	1000-3000	19.6		36-45	20.5
	3000-6000	28.7		46-55	15.4
	6000-10000	24.8		56-65	7.2
	10000-15000	11.4		"/> 66	2.8
	15000-20000	4.4	Education	Primary school	0.9
	"/>20000	3.8		Junior high school	5.3
	No answer	2.8		Senior high school	16.7
				College	28.3
				University	48.8

Source: consumer survey in September, 2008

Table 8.3 Socio-economic characteristics across treatments

In this survey, around 250 respondents are interviewed, but only 209 questionnaires are completely finished. The respondents are Beijing local citizens or people who have lived in Beijing longer than three years. Most respondents are female due to the fact that shopping is mainly done by housewives and in this survey, 69% of the correspondents are females while only 31% are males. The majority of the respondents are confined within the ages of 26-35, 19-25 and 36-45 years old. Additionally, the sample shows that the respondents have quite high levels of education. This is because Beijing is the capital city, where many famous universities are located. Hence in this survey, some 28% of the respondents answered as having college-level educational experience, and 48% of them responded as having attended universities or postgraduate schools. Nearly 80% of the respondents have household income less than 10000 yuan per month, about 29% of them have monthly household incomes of 3000-6000 yuan, and 25% have monthly household incomes of 6000-10000 yuan on average per month. Around 20% of the respondents have a monthly household income of more than 10000 yuan, while only less than 4% have a household income of less than 1000 yuan per month.

8.3.3 Research method

This study also applies the choice modelling (CM) technique in examining which attributes are significant determinants of the values people place on non-market goods, i.e., the traceability system. CM or stated preference (SP) that uses the attribute-based technique was first applied by Louviere et al (2000) [12], Louviere et al (1983) [13] and Adamowicz et al (1998) [14]. This technique originated in market research and transport literature, and has recently been applied to the valuation of non-market goods. In this survey, attributes and levels are used to create choice sets using a 3^3x6 orthogonal effects design which produced 36 choice sets and were divided into six versions. CM techniques require respondents to compare and select one option out of three in all the choice sets as shown in Table 8.4.

	Choice A	Choice B	Choice C
Farm and/or farmer	Information + Pictures	Information	I would buy my usual brand of milk
Veterinary medicine use	All medicine record	Without record	
Processing plantings	Information	Information + Pictures	
Price of 250ml milk	No change*	0.20 yuan more expensive	
I choose			

* No change means the basic milk price. The price of milk broad sold in China is 1.70 yuan per 250ml

Table 8.4 Choice and set on traceability system

The multinomial logit model is used to analyse the data. The options chosen by the respondents in the CM can be modelled in a random utility framework which can be expressed as the sum of the systematic component. The utility obtained by individual *i* from choosing alternative *j* in a choice set which can be expressed as:

$$U_{ij} = V_{ij} + \varepsilon_i \tag{8-1}$$

where V_{ij} denotes the observable portion of the utility and ε_{ij} indicates the error term. This study assumes that the utility for an option (i) depends on a vector of its observable attributes (Z) and a vector of the socio-economic characteristics of respondents (S) as:

$$U_{ij} = V_{ij}(Z_{ij}, S_{ij}) + \varepsilon_{ij}(Z_{ij}, S_{ij}) \tag{8-2}$$

Option j is chosen over alternative h of $U_{ij} > U_{ih}$. Probability if individual i choosing option j can be defined as follows:

$$\pi_{ij} = \Pr\left\{V_{ij} + \varepsilon_{ij} \geq V_{ih} + \varepsilon_{ih}; \forall h \in C\right\}, \tag{8-3}$$

where C_i is the choice set for individual i; V_{ij} is a conditional indirect utility function and has a linear form of:

$$V_{ij} = \beta_0 + \beta_1 X_1 + \beta_2 X_2 + \ldots\ldots + \beta_n X_n \tag{8-4}$$

where β_1 - β_n is the vector of coefficient attached to the vector of attributes X. While the socio-economic characteristics impact on attributes, the function has a form:

$$V_{ij} = ASC + \sum_{k=1}^{K} \beta_k Z_{ik} + \sum_{k=1}^{K} \sum_{h=1}^{H} \gamma_{kh} Z_{ik} S_{ih} \tag{8-5}$$

where β and γ is parameter, Z is the attribute associated with the alternative, S represents socio-economic characteristics.

The marginal value of a change within a single attribute can be represented as a ratio of coefficients as follows:

$$MWTP = \frac{\beta_{attribute}}{\beta_{price}} \tag{8-6}$$

Option C was coded as zero value and alternative specific constants were equal to 1 with either option A and B being selected [15]. In this study the software package LIMDEP 9.0 NLOGIT4.0 was used to estimate the multinomial logit model [16].

8.4 Results and discussion

8.4.1 Consumers' milk purchasing behaviour and reaction on food safety issues

The survey was carried out in Beijing in 2008. As one of the highest average milk consumption zones in China, consumers' milk purchasing behaviour was also examined. As the modern supply chain developed, supermarkets became the most popular place for people to buy daily goods. So people in Beijing are used to buying milk and other dairy products in supermarkets. Besides buying at the supermarket, milk delivery services are also common.

Source: consumer survey in September 2008

Figure 8.4 Main milk purchasing place

For many families, milk is consumed every day. Some people have milk at breakfast, while others have milk at night before sleeping, or both (about 250 ml -500ml per day). A few people consume milk as their main drink and drink milk very often (more than 500 ml per day). For old people and young children, milk, yogurt and other dairy products are the main source of calcium. Nearly 60% of families are used to consuming more than 250ml of milk per day per person, around 17% of the families are used to having more than 500ml of milk per day. Around 11% of the respondents rarely drink milk everyday (Figure. 8.5).

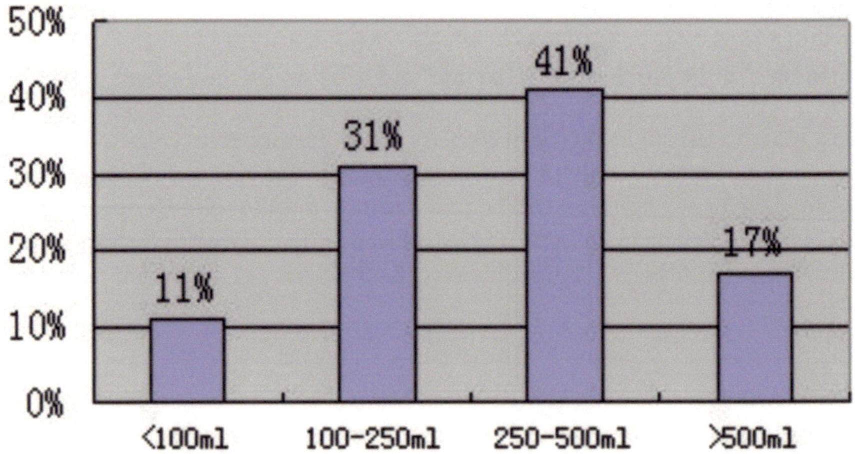

Source: consumer field survey in September, 2008

Figure 8.5 Milk consumption per household in a day

However, a few food safety incidents have negatively affected consumers' confidence on the food they consume, for example, the milk powder incident which happened in China in 2008. Before the incident happened, most people felt the milk was safe to consume, but after the incident, the situation totally changed [17] (Figure 8.6).

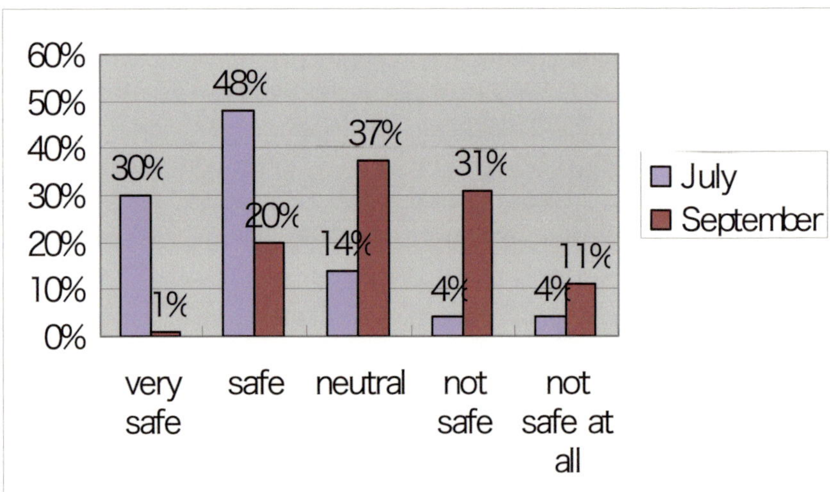

Source: field survey in July and September, 2008

Figure 8.6 Consumer's confidence on food safety in 2008

After the incident happened, respondents' views on safety become more important than price. Most respondents thought safety is much more important than milk price [9] (Figure 8.7).

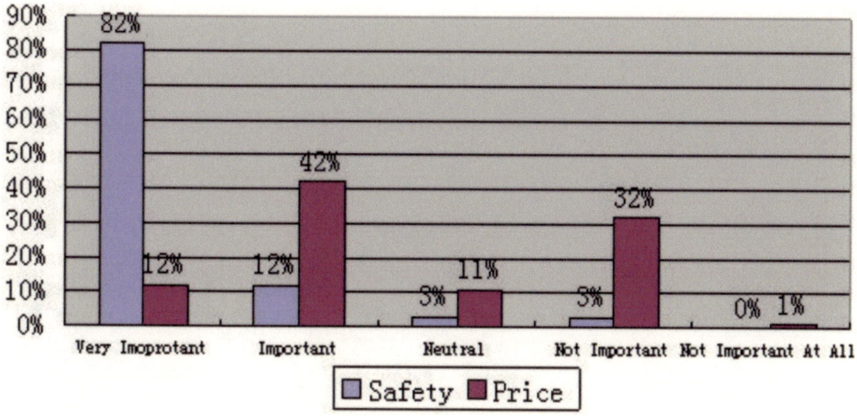

Source: field survey in July,2008

Figure 8.7 Consumer perceptions on safety and price

After the incident happened, most respondents wanted more heavy supervision of milk production, some people wanted more production information, while some people stopped purchasing milk altogether (Figure 8.8).

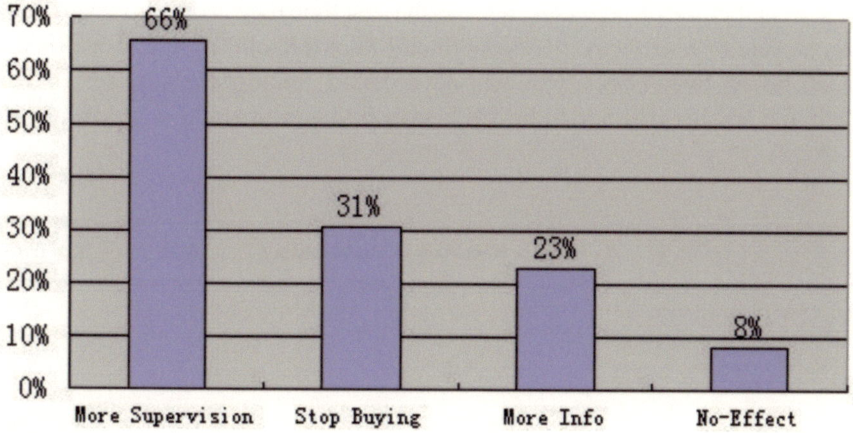

Source: consumer field survey in September, 2008

Figure 8.8 Consumers' requests on ensuring food safety

According to the survey conducted in July 2008, the respondents thought most milk safety problems happened on farms and in processing factories [16]. Very few respondents thought milk safety problems happened during transportation or at the wholesale stage (Figure 8.9).

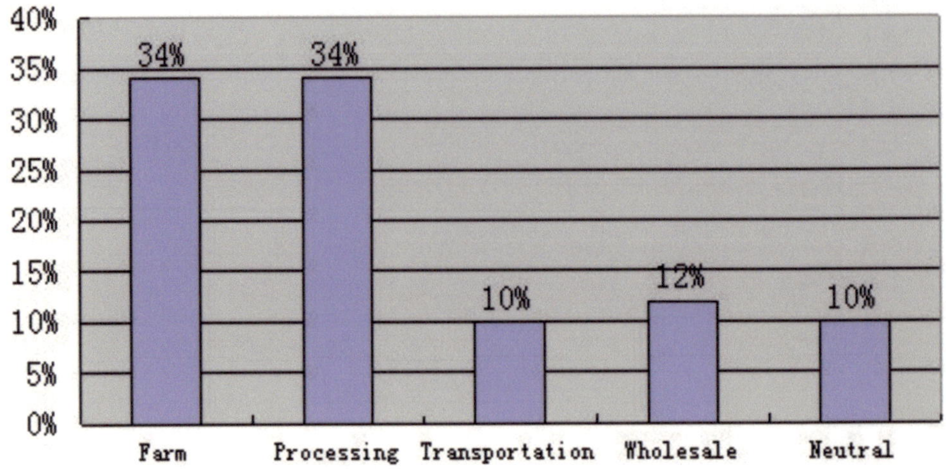

Source: field survey in July, 2008

Figure 8.9 Consumers' awareness on the most risky stage in food supply

8.4.2 Consumers' awareness of the traceability system

As a new system to ensure food safety, the traceability system is only known by a few people. Only around 42% of the respondents had heard about the food traceability system before, and for more than half of the respondents this was the first time they had heard about this system. However, around 87% of the respondents thought the traceability system was necessary to ensure food safety. Even people who had never heard about traceability system before the survey, still felt it was necessary and would be a way to ensure food safety and food quality (Table 8.5).

	Necessary	Unnecessary	Total	% of necessary
Heard about traceability system	78	11	89	87.6
Not heard about traceability system	104	16	120	86.6
Total	182	27	209	87.0
% of heard about traceability system	42.8	40.7	42.5	

Source: consumer survey in September, 2008

Table 8.5 Consumers' awareness of the traceability system

This result shows that most of the consumers accept the traceability system, whether or not they had heard of the traceability system. However, there were still around 13% of the respondents who did not approve of the traceability system. Distrusting the information was the main reason for this. Additionally, there were some respondents who did not care about the traceability system. They thought the traceability system was not perfect enough to ensure food safety and food quality. Only a few respondents viewed a possible higher price as the reason they would not choose a traceability system. At present, the traceability system is a hot topic in China; there is some research about consumers' attitude toward traceability systems, but on other products such as pork and beef. This result was similar to research on pork. According to Greene et al. (2002) [16], the reason for respondents not choosing pork with a traceability system was that people were worried about possible false information.

The reason traceability system is thought to be unnecessary	% in total responses
a. Higher price	3.8
b. Do not trust the information	55.5
c. Do not care about traceability system	37.0
d. Others	3.7

Source: Consumer survey in September, 2008

Table 8.6 Reasons for not choosing the traceability system

Price (yuan)*	Count	%	Price (yuan)	Count	%
1.70	56	31.1	3.60	1	0.5
1.90	62	34.4	4.00	0	0.0
2.20	25	13.8	4.60	3	1.6
2.50	20	11.1	5.20	2	1.1
2.80	8	4.4	6.00	0	0.0
3.20	2	1.1	"/> 6.00	1	0.5
Total	182	100.0			

Source: consumer survey in September, 2008

*Note: we set the basic price of milk as 1.70 yuan/250 ml with the assumed constant price

Table 8.7 Amounts of WTP for milk with the traceability system

Of 182 respondents, 30% would accept the traceability system if the price did not change. Around 70% of the respondents would accept an increased milk price with the traceability system. However, most of them only accepted a limited price increase. Only approximately

4% of the respondents were willing to pay a high price for the milk with a traceability system (three or four times more expensive than the original price of normal milk). This result showed that consumers were willing to bear part of the cost of a traceability system, but to a limited extent, and most of the cost should be borne by the government and producers. The government should take the responsibility to ensure the basic food safety (Table 8.7).

8.4.3 Consumers' willingness to pay for certified safety food

The respondents are mainly asked about their attitude toward the traceability system. Table 8.9 shows both the attribute variables and non-attribute variables used in choice modelling. Attribute variables are the information that the traceability system can provide to consumers, and non-attribute variables are mainly socio-economic information. The model estimated the price attributes which interacted with socio-economic characteristics and estimated how these socio-economic characteristics impact on price attribute. The variables are described as AGE*PRICE, GENDER*PRICE, and so on.

Table 8.10 showed the result of the estimation MWTP of information that the traceability system provides and the estimation of socio-economic characteristics which impact upon price attribute. According to Table 8.4, respondents preferred all the information except processing information with pictures. The reason may be that people are more familiar with these famous processing companies, they can get much information about these processing enterprises through many channels; however, most people do not have a clear idea about farm information, breeding information, feed information and animal medicine use which are all very important and affect food safety, especially feed use and animal medicine use.

With regard to farm information, consumers have a higher willingness to pay (WTP) on information with pictures than only information, that being 2.28 yuan and 2.03 yuan for 250ml of milk while the basic price of milk is 1.70 yuan. For consumers, the more information, the better. Additionally, consumers are willing to pay about 3.69 yuan for 250ml of milk with a traceability system which includes antibiotics records and only 2.95 yuan for all animal medicine records. This was a very high marginal willingness to pay, especially on antibiotics' usage. It was more than twice the original price. However, the survey was carried out right after the milk powder incident happened, so the result might have been biased and over estimated the real willingness to pay. Consumers are more concerned about animal medicine use, especially antibiotics' use. Processing factory information is viewed as least preferred, while processing factory information with pictures is not significant in the statistics. When asked about processing factory information, consumers have a lower WTP than other attributes at only 0.87 yuan, while the processing information with pictures is not significant in the statistics. The reason for this might be that consumers or the respondents already have enough information on processing factories, especially those famous brands compared with other information. They can get this kind of information through many channels such as news, the Internet, or see the factory for themselves. They might be more interested in some introductions to these processing factories, rather than pictures and people do not prefer the attribute of price through the coefficient.

Variables	Explanation	Codes
ASC	Alternatives specific constant	
FARM INF	Information of fairy farm described by words, the information might include the address, the contact information and some introduction of the farm.	1=FarmInf, 0=No-Information
FARM INF+PIC	Information of dairy farm with pictures. Besides information described by words, pictures might give consumers a direct image of the farm. For farmers, once their pictures can be found by the public, they feel they have the responsibility to provide safe food.	1=FarmInf+Pic, 0=No-Information
ANTIBIOTIC RECORD	Antibiotic record. The most important medicine used for cows.	1=Antibiotics, 0=No-Record
ALL MEDICINE RECORD	Entire animal medical record. Including antibiotic usage record and other medicine use records	1=All Record, 0=No-Record
PROCESSING INF	Processing factory information described by words, the information might include the address, the contact, processing method and some detail by way of introduction of the processing factory	1=ProcessingInf, 0=No-Information
PROCESSING INF+PIC	Processing information with pictures. Besides information described by words, pictures might give consumers a direct image of the processing factories.	1=ProcessingInf+Pic, 0=No-Information

Table 8.8 Explanation of attribute and non-attribute variables in the choice models

Variables	Explanation	Codes
GENDER	Respondent sex	0=Male; 1=Female
AGE	Respondent age	1=<=18; 2=19-25; 3=26-35; 4=36-45; 5=46-55; 6=56-65; 7="/>=66
EDU	Respondent Educational Level	1=Primary; 2=Secondary; 3=High School/ College; 4=Technical/Vocational; 5=University
FAMILYNO	No. of members in the household	Number of members
KID	No. of children in the household	Number of kids
OLDPPL	No. of old people in the household	Number of old people
INCOME	Total income of per household (yuan)	1=<1000; 2=1000-3000; 3=3000-6000; 4=6000-10000; 5=10000-15000; 6=15000-20000; 7="/>20000

Table 8.9 Explanation non-attribute variables in Choice models

Table 8.10 also shows the results of the impacts of socio-economic characteristics on price attribute (WTP). Only AGE*PRICE and EDU*PRICE are in 1% significant, INCOME*PRICE is in 10% significant. Other variables are not significant in the statistics. AGE*PRICE is negative, young people find it easier to accept the traceability system and are more willing to pay for a traceability system. EDU*PRICE is positive, higher educated people find it easier to accept the traceability system and have higher WTP. INCOME*PRICE is positive, higher income people are willing to pay more money, but not very much. This may imply that income only impacts a little on WTP for the traceability system. No matter their level of income, people are concerned about food safety and the traceability system, and they need safe food whether rich or poor.

Variable	Estimation of information from traceability system	Estimation of socio-economic characteristics impact on price	MWTP
ASC#	-0.674***	-0.443***	-2.07
FARM INF	0.659***	0.654***	2.03
FARM INF+PIC	0.741***	0.735***	2.28
ANTIBIOTICS RECORD	1.202***	1.206***	3.69
ALL MEDICINE RECORD	0.961***	0.960***	2.95
PROCESSING INF	0.282*	0.274*	0.87
PROCESSING INF+PIC	0.127	0.121	0.39
PRICE	-0.325***	-0.312***	
GENDER*PRICE		-0.0023	
AGE*PRICE		-0.029***	
EDU*PRICE		0.036***	
FAMILYNO*PRICE		-0.247	
KID*PRICE		0.022	
OLDPPL*PRICE		-0.001	
INCOME*PRICE		0.002*	
Rho-square	0.265	0.287	
Adjusted rho-square	0.262	0.277	
Number of observations	1254	1254	

Source: consumer field survey 2008

means alternative specific constants,

***, **and * denote statistically significant at 1%, 5% and 10%, respectively

Table 8.10 Estimation of MWTP

8.5 Conclusions and recommendations

The results suggest that consumers in Beijing are used to consuming milk in daily life. However, the food safety incidents which occurred have negatively affected consumers' confidence in the food they consume. The results also suggest that the traceability system is still a new thing to consumers in Beijing. They have very little knowledge of the traceability system, but they have a positive attitude toward the traceability system. More than half (57.2%) of the respondents had never heard of the traceability system before the survey, most (87.6%) consumers thought the food traceability system is necessary to ensure the food safety and avoid information asymmetry, although they are only willing to pay a limited amount for the traceability system.

Although respondents thought it is necessary to have traceability system to ensure food safety, they also worry about whether or not the information would be trustworthy. Most consumers could accept a traceability system and are willing to afford a small part of the cost. Therefore, doubts about the veracity of the information that a traceability system provides is the biggest problem at present. Powerful supervision is important to make people trust the information. People's behaviour is hard to control, but good supervision can reduce the risk of providing false information.

Consumers are used to buying milk and dairy products in supermarkets, they believe the modern food supply chain can ensure food safety more than the old tradition way. Milk has become a normal part of the diet for the people in Beijing and they are used to drinking milk every day.

Consumers are concerned about the information on animal medicine use, especially antibiotics, and are willing to pay more for the information on this. Additionally, people also care about the farm information and they thought the more the information, the better. So providing this information might increase consumers' confidence in the food they consume. These younger, higher educated and higher income people find it easier to accept the traceability system and are willing to pay extra money for the traceability system. But income is not a strong factor affecting willingness to pay.

As most dairy farmers in China are small farmers at present, it is still difficult to carry out a traceability system on dairy farms. However, according to the China Dairy Yearbook, the occupation of small farms is decreasing while the occupation of large farms is increasing, and therefore recording information and tracing back will be the trend in the modern dairy industry.

References

[1] Aizaki H. Nanseki T., Zhou H. Japanese consumer preferences for milk certified as good agricultural practice. Animal Science Journal 2012; doi: 10.1111/j.1740-0929.2012.01043.x.

[2] Nanseki T., Yokoyama K. JAPAN: Improving Food Safety amongst Food Operators, Ian G. Smith and Anthony Furness Ed. Food Traceability Around the World, Vicarage Publications Ltd, England 2008; Vol.1: 46-65.

[3] Borst P., Akkermans, H., Top, J. Engineering ontologies. International Journal of Human-Computer Studies 1997; 46(2-3): 365-406.

[4] Gordijin J., Akkermans H. Designing and evaluating e-business models. IEEE Intelligent System 2001; 16(4): 11-17.

[5] Sahin E., Dallery G., Performance evaluation of a traceability system, Proceedings of IEEE International Conference on Systems, Man and Cybernetics 2002; (3): 210-218.

[6] Dickinson D., Bailey D. Meat Traceability: Are US consumers willing to pay for it? Journal of Agricultural and Resource Economics 2002; 27: 348–364.

[7] Hobbs J. et al. Traceability in the Canadian Red Meat Market Sector: Do Consumers Care? Canadian Journal of Agricultural Economics 2005; 53:47–65.

[8] Mu J. Consumers' willingness to pay for a traceability label on poultry meat - which will be effective to calibrate hypothetical bias in CVM: Cheap Talk or Uncertainty Adjustment. Thesis of Master Degree in the People's University of China 2006.(in Chinese)

[9] Nanseki T., Xu Y., Zeng Y. Feasibility study for comparison of food risk perception in Japan and China, Proceedings of Annual Symposium of the Farm Management Society of Japan 2008: 222-223 (in Japanese).

[10] Official Website of the Beijing Government. Capital planning of main functional regions: http://zhengwu.beijing.gov.cn/ghxx/qtgh/t1240927.htm (Accessed on 19 Oct., 2012)

[11] China Ministry of Agriculture.China Statistical Yearbook 2012. Shenyang: Liaoning Education Press, 2012.5.

[12] Louviere J., Hensher D., Swait J. Stated Choice Methods: Analysis and Application. Cambridge University Press, 2000.

[13] Louviere J., Woodworth G. Design and Analysis of Simulated Consumer Choice or Allocation Experiments: An Approach Based on Aggregate Data, Journal of Marketing Research 1983; 20: 350-367.

[14] Adamowicz W., Louviere J., Swait J. Introduction to Attribute Based Stated Choice Methods. Report to NOAA Resource Valuation Branch, Damage Assessment Centre, 1998.

[15] Bateman I. et al. Economic Valuation with Stated Preference Technique: A Manual. Edward Elgar. United Kingdom; 2002: 458.

[16] Greene W. LIMDEP version 9.0 Econometric Modeling Guide Vol.2: Economic Software Inc, NY, USA, 2002.

[17] Min S., Liu L., Wang Z., Nanseki T. Consumers' Attitudes to Food Traceability System in China: Evidences from the Pork Market in Beijing, Journal of the Faculty of Agriculture, Kyushu University 2008; 54(1): 569-574.

Chapter 9

Awareness Comparison Between Farmers and Consumers

Hui Zhou and Teruaki Nanseki

9.1 Introduction

A number of food safety problems within China have negatively affected consumers' confidence in the safety of food products [1-2]. Information asymmetry between the consumers and the producers, especially the farmers, is considered one of the most important reasons why consumers are worrying about food safety in China [3-4]. In the foregoing chapter, perceptions of both the farmers as producers of food and consumers are surveyed and analysed. Further analyses and understanding of attitudes towards food safety from both sides are important for ensuring food safety.

In this chapter, we intend to compare the attitudes towards food safety problems and agricultural risk between consumers and producers (farmers), and to estimate the factors affecting farmers' knowledge of certified safe food.

9.2 Data collection

The analysis of this section is based on two field surveys. The first one is a survey of farmers from six provincial regions as introduced in Chapter 3. The other is a consumer survey held in Beijing, 2008. Consumers' attitudes towards rice, vegetables, meat and milk are included in the questionnaire. A total of 186 samples were taken in July and 209 samples in September. The consumers' survey conducted in July is mainly used in this case study. The respondents from this survey include both a supermarket survey and a home-visiting survey.

Table 9.1 shows the demographic characteristics of the consumers and farmers from the two surveys. In addition to the basic information of the respondents in the surveys, we can conclude that the consumers have higher levels of education and cash income, from comparing data provided in the two columns. Among the consumer, more than 53% of them attended colleges and high schools, while the ratio in the sampled farmers is less than 5%. Meanwhile, more than 60% of the consumers have annual income over 5000 yuan, while the same proportion among farmers accounts for only 21.2%.

Consumers of Beijing			**Farmers from six provincial regions**		
Gender	Male	45.2%	Gender	Male	81.9%
	Female	51.3%		Female	18.1%
	No answer	3.5%			
Marriage	married	60.2%	Marriage	married	N/A
	single	37.4%		single	
	Others	2.4%		Others	
Age	"/>18	2.1%	Age	20-30	4.2%
	18~29	41.2%		30-40	18.6%
	30~49	33.4%		40-50	35.2%
	50~64	16.2%		50-60	28.6%
	Over 65	7.1%		Over 60	13.4%
Edu	Illiterate	1.2%	Edu	Illiterate	17.1%
	Elementary	5.2%		Elementary	26.2%
	Middle School	13.3%		Middle School	43.3%
	High School	27.1%		High School	17.8%
	Undergraduate	48.0%		Undergraduate	3.2%
	Graduate	5.2%		No respond	5.4%
Monthly Income (yuan)	< 1,000	2.2%	Annual Income		
	1,000~2,000	9.2%		< 10,000	12.1%
	2,000~3,000	12.3%		10,000-30,000	33.7%
	3,000~4,000	14.1%		30,000-50,000	33.0%
	4,000~6,000	22.1%		50,000<	21.2%
	6,000~8,000	18.1%		No respond	1.0%
	8,000<	22.0%			

Source: field survey in 2011, consumer survey in 2008 (July)

Table 9.1 Demographic characteristics of the consumers and farmers

9.3 Research method

Both description analysis and the economics model are used in this research. Description analysis is used to examine the difference of attitude towards food safety problems between consumers and farmers, while a binary logit regression model is adopted to examine the factors affecting farmers' awareness of food safety-related certification. Previous research, such as that by Carlson, posit that some social-economic characteristics, such as education level, income level and others, positively affect farmers' behaviour on protecting the agro-environment [5].

The model is shown as:

$$y=\beta_0+\beta_1x_1+\beta_2x_2+\ldots+\beta_kx_k+u \tag{9-1}$$

while the distribution function of y is:

$$f(y)=P_y\ (1-P)_{1-y}(y=0,\ 1) \tag{9-2}$$

where farmers answer the question about food-related certification correctly, y=1; otherwise, y=0; x are a series variables effecting farmers' knowledge of food-related certification (shown below); and while u is the random error. The model is calculated by using SPSS 13.0 for Windows.

	Category	Description
Independence variables	Age	Age of the household head
	Edu	Education level of household head
	Income	Annual income of a household
	Location	Beijing, Hebei and Shandong=1(North); Shanghai Jiangsu and Zhejiang=0(South)
	Off-farm work	Doing off-farm work=1, otherwise=0
Dependence Variables	Knowledge on food safety-related certification	Answer the question about food safety-related certification correctly=1, otherwise=0

Table 9.2 Description of variables

9.4 Results and discussion

9.4.1 The attitude towards the food safety situation

Figure 9.1 shows the different attitudes towards the food safety situation between consumers and farmers. In 2008, a serious food safety scandal, the Sanlu incident, happened in the dairy industry in China which destroyed consumers' confidence in food safety. Before the scandal, consumers showed high confidence in food safety, even higher than farmers. Once the food safety problems occurred, they lost faith very quickly and consumers have become more sensitive to food safety than farmers. As shown in Figure 9.1, 30% and 48% of the consumers surveyed in July 2008 thought that the situation of food supply as *very safe* and *safe*, while only 1% and 20% of the consumers appraised it as *very safe* and *safe*, when they were interviewed in September of 2008 after the occurrence of the Sanlu incident. In July 2008, most of the consumers (48%) chose the appraisal of *safe*, while the largest proportion (37%) appraised the food supply situation as neutral in the survey in September 2008.

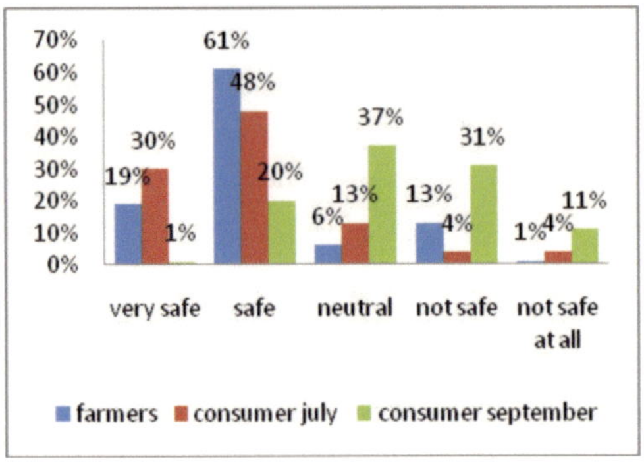

Source: field survey in 2011, consumer survey in 2008

Figure 9.1 Consumers and farmers' attitude towards food safety

9.4.2 Risk of food safety in different stages

Figure 9.2 shows the different opinion among consumers and farmers about at which stage of the supply chain food safety issues are most likely to happen. Consumers' opinion on rice is examined here. Most of the sampled farmers grow rice, corn or wheat crops, thus consumers and farmers have different ideas on which stage the food safety issues are most likely to happen. Most of the farmers (50%) believe that the agro-products are most likely to be polluted at the production stage, while consumers chose the most risky as processing accounting for the largest proportion at 50%.

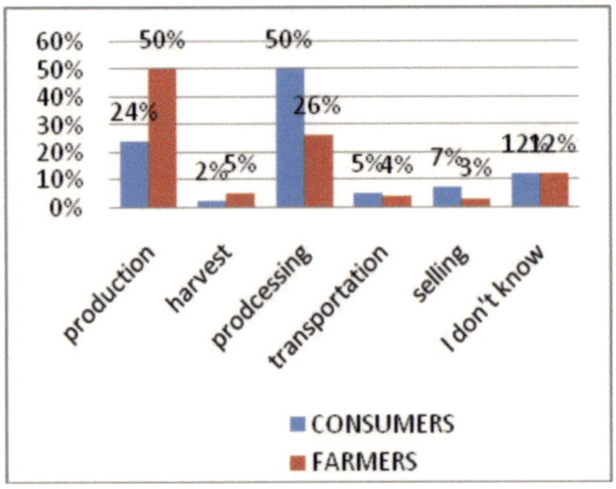

Source: field survey in 2011, consumer survey in 2008

Figure 9.2 Risk of grain be polluted in different stages

9.4.3 Information source of food safety and related news

Table 9.3 and table 9.4 shows the main information source of food safety and related news for both consumers and farmers. TV and radio are the most common ways for both consumers and farmers to get information relating to food safety, while consumers have more channels to get this kind of information than farmers. They both trust the information provided by government more than other resources.

Category	Consumers	Farmers
TV and radio	144	237
Newspapers and magazines	100	35
Friends, relatives	78	69
Internet	56	N/A
Advertisement	29	N/A
Label on food package	90	29
Doctor and specialist	33	40
Others	8	4

Source: field survey in 2011 and consumer survey in 2008

Table 9.3 Information channels of both consumers and farmers

Category	Consumers	Farmers
Government	162	303
Doctors and researchers	131	67
Friends and relatives	97	49
NPO	81	N/A
Companies	54	N/A
Others	17	58

Source: field survey in 2011 and consumer survey in 2008

Table 9.4 Most trusted information sources to consumers and farmers

9.4.4 Farmers' awareness of food safety certifications

Most of the farmers surveyed do not have a clear idea about food safety-related certifications such as green food, organic food and nuisance-free food. Many farmers believe that green food is the best among those three certifications. Only 23% of the respondents are correct in identifying the right order of green food, organic food and nuisance-free food. Hence we can conclude that farmers' awareness of certificated food is still low.

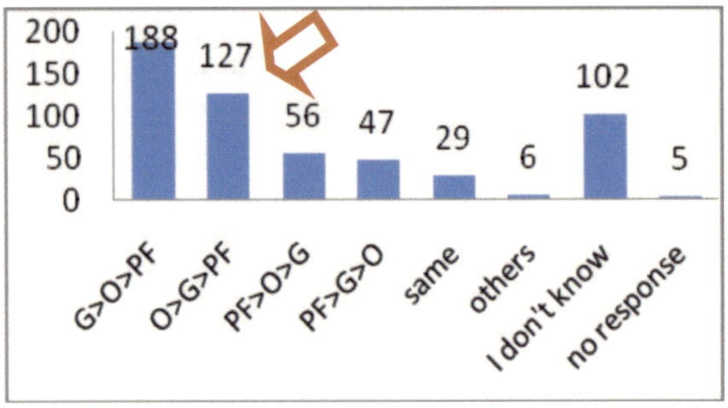

Note: G= Green Food; O= Organic Food; PF= Nuisance-free Food. The correct order is: O>G>PF
Source: field survey in 2011

Figure 9.3 Farmers' understanding of food safety-related certifications

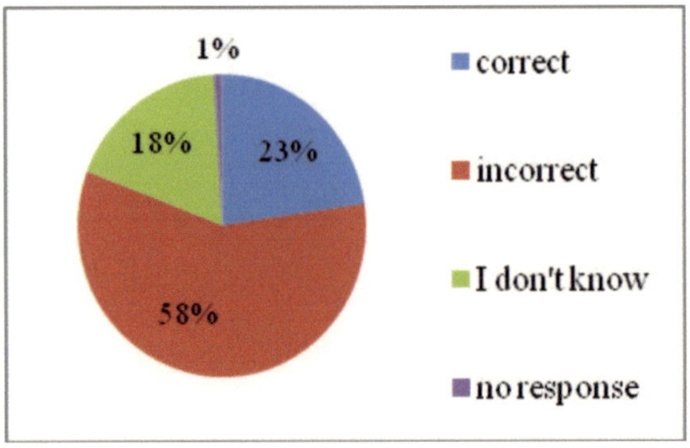

Source: filed survey in 2011

Figure 9.4 Farmers' knowledge of food safety-related certifications

9.4.5 Factor affecting farmers' awareness of food safety-related certifications

A binary logistic model is adopted to examine factors influencing farmers' knowledge of food safety-related certification. The relevant statistics of the model are shown in Table 9.5. The significant values of F and t (p-value < 0.1) indicate the good fitness level of this model[1].

	B	S.E.	Wald	df	Sig.
Age	-0.023**	0.011	3.977	1	0.046
Edu	-0.146	0.137	1.130	1	0.288
Off-work	-0.470*	0.251	3.506	1	0.061
Income	0.189	0.132	2.047	1	0.153
Location	-0.576**	0.257	5.003	1	0.025
Constant	0.166	0.938	0.032	1	0.859

Chi-square=18.146***

Nagelkerke R Square=0.232

When farmers answer about food safety related certification correctly, y=1, otherwise, y=0

Note: ***, ** and * represent statistical significance in level of 1%, 5% and 10%, respectively.

Source: field survey in 2011.

Software: SPSS 13.0 for windows.

Table 9.5 Effects of the factors on farmers' awareness

Only age, off-farm work and location are significant, while others do not strongly affect farmers' awareness of food related certifications. Considering those food safety related certifications are new to farmers, those young farmers are more familiar with those certifications than older ones and more easily to understand and accept those new things. Farmers do less off-farm work may focus more on agriculture. They may have more knowledge and pay more attentions on those new things.

In this case study, sample farmers from south part know more about food safety related certifications than farmers from north part. In this survey, Jiangsu Province, Zhejiang Province and Shanghai are defined as south part; these three provinces and city are viewed as developed areas in China. The economic situation in these areas is better than the north part of current study site. It is a possible reason the sample farmers are more familiar with this kind of knowledge.

1 The R2 value of 0.406 should not be used to judge the fitness of a model. The fact that R2 never decreases when any variable is added to a regression makes it a poor tool for deciding whether one or several variables should be added to a model. Low R2s in regression equations are not uncommon, especially for cross-sectional analysis. Thus, using R2 as the main gauge of success for an econometric analysis can lead to difficulties [6].

9.5 Conclusion and policy recommendations

The attitude towards food safety situation is much different between consumers and farmers. The information asymmetry caused the attitude or the awareness of food safety situation is different between consumers and farmers.

Farmers and consumers have different ideas about at which stage food safety problems are most likely to occur. But for information on food safety and related issues, both consumers and farmers share similar idea. Both consumers and farmers trust the information provided by government.

Farmers have less knowledge about food safety certification and related topics. Age, off-farm work and location are significant variables effecting farmers' knowledge on food safety-related certification. Younger farmers, farmers who do less off-farm work and farmers from the southern part of China are more familiar with the knowledge. However, income level and education level have no strong relationship with farmers' knowledge on those certifications.

An information platform is necessary to connect both consumers and producers, and third party food safety certifications are needed to reduce the information asymmetry, however, a supervision system is necessary to keep the third party certification trustworthy.

Although food safety-related certifications and systems are supposed to be useful in ensuring food safety and reducing the information asymmetry, however, farmers have less knowledge about them. The government should therefore provide more programmes to help farmers gain more knowledge about them.

There are still some limitations in this research. In the model, agricultural experience should be one of the variables, education level may not be significant in influencing farmers' knowledge on food safety-related certifications; while agricultural experience, and whether or not farmers get extension services are more important for them to understand and accept new concepts such as organic food and others. However, data on this in the current survey is not available. In the future research, some new variables should be included and the model should be completed.

References

[1] Song M., Gao X., Nanseki T. Reducing food safety risk: agricultural practices in China. Journal of the Faculty of Agriculture, Kyushu University 2010; 55(1): 569-57.

[2] Xu Y., Nanseki T., Zeng Y. Food safety issues and consumer perception in China. Nanseki T. [Ed] Food risk and safety in East Asia, Agriculture and forestry statistics publishing Inc 2010; 101-119 (in Japanese).

[3] Song M., Liu L., Wang Z., Nanseki T. Consumers' attitude to food traceability system in China: Evidences from the pork market in Beijing, Journal of the Faculty of Agriculture, Kyushu University 2008; 53(2): 569-574.

[4] Sun Q. Analysis on the determinants of production of safety agricultural products by farms, Chinese Journal of Food and Nutrition in China 2008 (1); 15-17.

[5] Carlson J., Dillman D. Early adopters and non-users of No-till in the Pacific Northwest. Report prepared for the soil conservation service by the department of rural sociology, Washington State University and Department of Agricultural 1985.

[6] Wooldridge J. Introductory Econometrics: A Modern Approach (2nd Edition). South-Western Thomson Learning, Mason, 2003: 41, 81.

Chapter 10

Risk Governance System for Food Safety in Japan and China

Teruaki Nanseki and Min Song

10.1 Introduction

Consumers' concerns over food safety are always crucial issues in the world. These concerns are a result of various events relating to food safety. Some notable examples of these events are bovine spongiform encephalopathy (BSE), more commonly known as mad cow disease; residues of agro-chemicals in agricultural products and food contamination by environmental pollution. Food safety issue is more complicated because of world trade of food. As an example, Japan imports a huge amount of food from foreign countries including China, which is one of the biggest exporting countries. Therefore, it is necessary to improve the safety of not only domestic food, but also imported food for Japanese consumers. In other words, the food traceability system and risk management at the farm level are crucial for food safety in China as well as in Japan.

This chapter aims to establish an academic basis for the development of a risk governance system for food safety in East Asia using the cases of Japan and China. First, the histories of food safety policy in Japan and China are briefly reviewed. Second, the current statuses of the food traceability system in both countries are clarified. Third, consumer perception on food safety is analysed from various aspects. Fourth, the current statuses of risk management at the farm level (e.g., good agricultural practices or GAP) in both countries are described. Finally, concluding remarks is given for further research.

10.2 Data source and survey method

Various survey data, including several of the authors' original surveys, and government statistics are used for analysis in this chapter (Table 10.1). The authors' original survey on consumer consciousness of food safety was done in Japan and China, 2008 [1]. The respondents of the preliminary surveys are 297 in total in China. The survey in Japan was an indoor group investigation using a survey slip. The survey in China was an individual interview. The authors' original survey on consumer consciousness on pork and milk traceability was undertaken in China [2].

An Internet survey on food traceability was conducted in Japan for 1059 registration monitors by a Japanese survey company (see goo Research (2007) [3]). A survey for 4150 food retailers was undertaken in Japan using the mail survey method by the Ministry of Agriculture, Forestry and Fisheries [4]. A nationwide survey on public opinion on food and agriculture was undertaken in Japan of 5000 people using the individual interview method by the Cabinet Office (2008) [5].

Reference or investigator	Name of survey	Survey respondents and research method	Sample information (see note)	Research date and remarks
The authors' original survey (2008) [2]	Consumer consciousness survey concerning food safety (preliminary survey, in Japanese and Chinese)	Consumers in Fukuoka, Japan and Beijing, China. Indoor group investigation using survey slips in Japan, and individual interviews and survey slips in China.	a and b: 111 in Japan a and b: 124 in China	July to October 2008
The authors' original survey (2008)	Consumer consciousness on milk traceability (in Chinese)	Consumers in Beijing, China. Individual interviews and survey slips.	a and b: 209	September 2008
The authors' original survey (2008) [1]	Consumer consciousness on pork traceability (in Chinese)	Consumers in Beijing, China. Individual interviews and survey slips.	a: 401, b: 388(valid), c: 96.8%	December 2007 to January 2008
The Cabinet Office (2008) [5]	Public opinion poll concerning role of food, agriculture, and farm village	20-year-old residents, individual interviews, two-step stratification random sampling	a: 5000, b: 3144, c: 62.9%	September 2008
The Ministry of Agriculture, Forestry and Fisheries (2008) [4]	Food industrial trend investigation (in Japanese)	Food retailers, mail surveys	a: 4150, b: 2085, c: 50.2%	As of January 1, 2008, January to February 2008
goo Research (2007) [3]	General results on residents' opinions regarding food traceability (in Japanese)	goo Research registration monitors, closed-door type Internet survey	a: unknown, b: 1059 (valid response)	September 2007

Note: a: number of investigations, b: number of respondents, c: response rate

Table 10.1 Major survey characteristics in Japan and China

10.3 Food safety policy in Japan and China

10.3.1 Food safety policy in Japan

In April 2002, the MAFF announced a restructuring plan for food and agriculture in order to deal with the problem of food safety. In May 2003, the Food Safety Basic Law was enacted (Figure 10.1) to promote food safety. The Food Safety Commission was established in July of the same year based on the law, and was tasked to scientifically and objectively assess the effect of food on health, and to make recommendations on food policy. In July 2003, due to changes in policy, the MAFF was restructured, and to promote risk management and to administer the changes for consumers, a safety bureau and a consumption bureau were newly established.

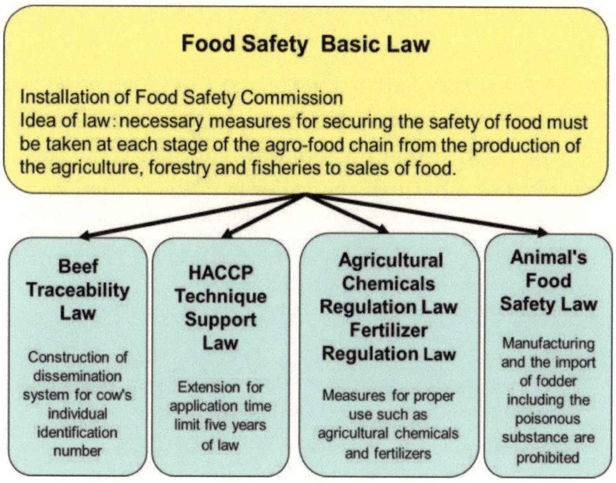

Figure 10.1 Basic Food Safety Law and related laws in Japan

Subsequently, the Beef Traceability Law was enacted, and related laws, namely the Animal's Food Safety Law, Agricultural Chemicals Regulation Law, Fertilizer Control Law, Hazard Analysis and Critical Control Points (HACCP) Technical Support Law, and others were revised.

10.3.2 Food safety policy in China

Since 2000, China has adopted numerous measures and programs regarding domestic food safety and international trade, in order to introduce, extend, encourage, and even mandate a traceability system in the food supply chain. In the legislation before 2001, there are specific laws or regulations concerning food safety, but little referring to traceability. However, in recent years, a traceability of food system gradually evolved into the necessary policy options. According to the law of the People's Republic of China on agricultural products' safety and quality issued in 2006, all agricultural enterprises must establish an authentic production recording system, from which data should be kept for at least two years; otherwise, transgressors will be penalized with not more than 5000 yuan. In addition, individual producers are also encouraged to keep records on their own production. Such actions are considered as the rudiments of the food traceability system in China [1].

10.4 Food traceability system in Japan and China

10.4.1 Food traceability system in Japan

The introduction ratio is high among meat retailers and large-scale retailers, but is low among vegetable and fruit retailers, and small-scale retailers. The main reasons for non-introduction are as follows. (1) There is no budget for the equipment and staff required to introduce and implement the system (41.2%). (2) The introduction method is not known (39.4%). (3) Neither the suppliers nor the customers request it. In 2007, 31.6% of the investigated food retail operators can identify the producer of all products, and 40.3% can specify the producer of some products, representing a rise of 7.7 percentage points from 2005. The main reasons for non-specification are as follows. (1) Information about the supplier is not recorded and kept by the retailer (47.4%). (2) Information about the supplier is not recorded and kept with the trader (40.3%).

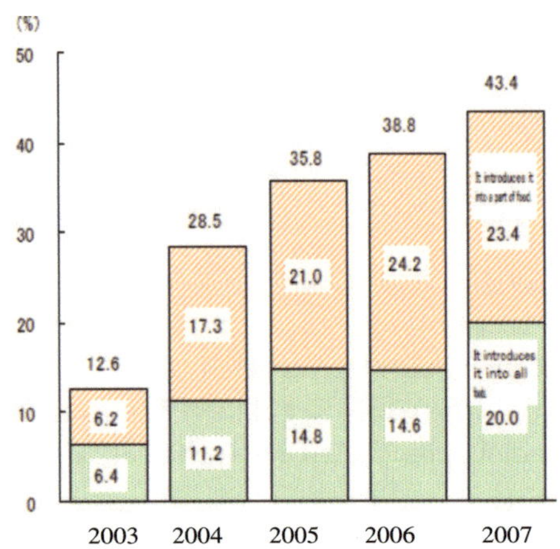

Source: Ministry of Agriculture, Forestry and Fisheries (2008) [4]

Figure 10.2 Increase in traceability system adoption in the food industry in Japan

By law, the traceability system is obligated for beef and rice. However, there is no law obligation for other types of food. Moreover, vegetables and fruits are being produced by many small-scale producers. Therefore, the standardization of the commodity is difficult, and the marketing channel complex. In addition, even if the cost is subsidized by the government for an individual producer, wholesaler, and retailer, the advantage of the traceability system introduction is not clear.

10.4.2 Consumer's perception of the system in Japan

This section presents the consumer evaluation of the food traceability system in Japan based on a survey by a Japanese research survey company [3]. In 2007, 19.1% of the respondents understood the content of the traceability system, and 30.7% know only the name. However, only 7.1% had experienced actually examining the traceability of food. Although the consumers want to buy safe food, evaluating food safety from information provided by the producers and the distributors is difficult. Some 31.0% of the respondents could not evaluate the information provided by the system itself. Some 23.3% did not trust the information provided by the system based on their negative experiences in the past. Therefore, the consumers do not understand the importance of the system, and consequently, the resulting additional costs that would be passed onto them for implementing the system. The survey showed that 26.7% of the consumers do not want to pay any additional cost for traceability information, and 71.6% do not want to pay any additional cost of more than 4% of the price of food for traceability information (Figure 10.2). The most important purpose of the traceability system is to enable a prompt recalling of polluted commodities should a contamination occur. Traceability information is important to tracing the commodity forward and back. However, product recalls are done by distributors and producers, and not by consumers. In this sense, consumers do not need traceability information, which explains the survey results.

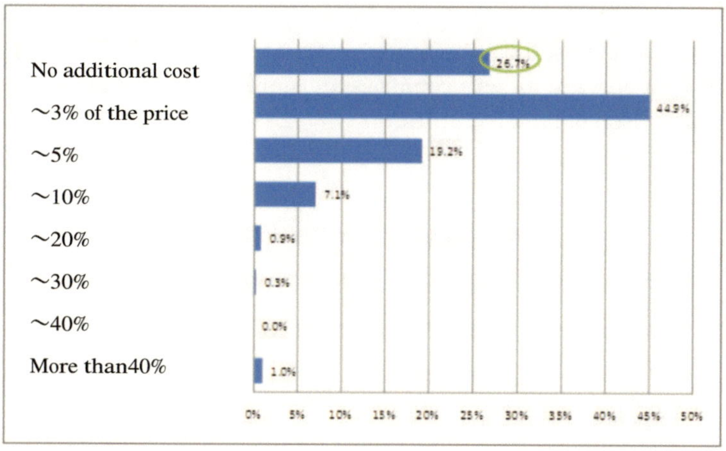

Source: goo Research (2008) [3]

Figure 10.3 Additional payment for food traceability in Japan

The respondents who had used the system questioned the trial use of the system. According to them, the system should be used (1) because doing so is publicized in the news (58.5%), (2) to confirm production region information (32.1%), (3) to confirm producer information (24.5%), (4) to confirm traceability of the purchased food (13.2%), and to confirm the agro-chemical history of the purchased food. The survey also showed that the consumer is more interested in information on the production rather than the distribution process.

10.4.3 Food traceability system in China

As early as 1992, the "Green Food" program was launched by the Ministry of Agriculture (MOA). In April 2001, the "Action Plan for Nuisance-Free Agricultural Products" was formally carried out by the MOA. These actions in part aimed to install simple traceability systems. Up to now, even though there is no formal, unified, and standard food traceability system implemented for the whole market, increasingly more wholesale markets and agricultural product suppliers are building diverse forms of specific food safety systems, using measures such as appointing and specifying suppliers, checking receipts and tickets, and detailed market recording. A survey covering 1329 city markets and 1108 rural markets performed in 2006 by the Ministry of Commerce of China showed that 53.7% of city markets, 32% of supermarkets, 80.4% of wholesale markets of agricultural products, and 70.7% of retail markets have introduced the above initial measures to foster traceability (Ministry of Commerce of China, 2006). More than half of the frozen food available in 2008 could be traced back to its origin.

10.4.4 Consumer recognition of the system in China

The authors' original consumer survey was conducted in 2007 on the pork market of Beijing [1]. The results showed that only 1.5% of the respondents are very familiar with the system and 12.1% rather familiar (Figure 10.4). Some 26.2% have moderate knowledge and 32.2% slight knowledge. Further, 27.8% know nothing about the system. Several surveys by the authors in Japan and China indicated that the recognition rate of the food traceability by consumers in China is almost the same or higher than that in Japan. Although the objectives and methods for the surveys in both countries are similar, a further survey is needed for a more precise comparison of both countries.

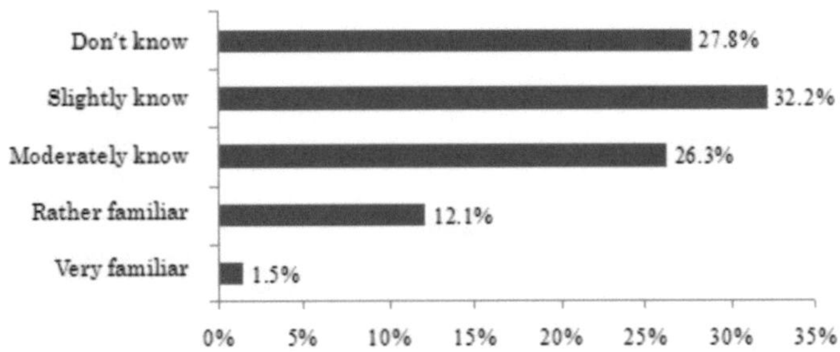

Source: Min et al. (2008) [1]

Figure 10.4 Consumers' knowledge level of the food traceability system in China

The respondents stated that they would purchase pork with traceability information because it assures the pork quality. Surprisingly, however, they do not care about the details of the food traceability system. On the other hand, the respondents stated that they would not purchase pork with traceability information for the following reasons: lack of trust in the authenticity of the information in the system (42.1%), lack of knowledge about the system

(26.7%), and potentially higher prices caused by the system (23.7%). Table 10.2 presents further results on this survey. The surveys indicated that in both countries, many consumers do not trust the information provided by food traceability. This lack of trust is therefore a key issue for food traceability.

Reason for choosing pork with traceability information	N	%
It can help me distinguish among various pork products based on production type	47	12.3
It can assure pork quality after knowing the production process	294	77.2
It can help me know more about producers and encourage my loyalty to them	21	5.5
It can help me exchange information with producers	19	5.0
Other	0	0.0
Reason for do not choosing pork with traceability	**N**	**%**
Higher price	63	23.7
Knowing little about it	71	26.7
Not necessary	19	7.1
Possible unauthentic information	112	42.1
Other	1	0.4

Source: Min et al. (2008) [1]

Table 10.2 Reasons for choosing pork with or without traceability information in China

10.5 Consumer's perception on food safety in Japan and China

The authors' original survey in Japan showed that consumers recognize several factors affecting food safety. Some 72.1% recognize the residues of agro-chemicals in rice, and 58.6% in vegetables. For rice, consumers also recognize heavy-metal contamination (53.2%), as well as impurities (36.0%), quality degradation (34.2%), genetic modification (27.9%), and food additives (18.9%). For vegetables, they also recognize bacillus contamination and rotting (23.4%), as well as genetic modification (9.9%) (Figure 10.5).

Similarly, the authors' survey in China showed that consumers recognize several factors that decrease food safety. Some 50.0% of the investigated consumers recognize agro-chemical residues in vegetables and 34.5% in rice (Figure 10.6). This result does not mean that agro-chemical residues are the biggest factor in food contamination statistically, but rather an indication of the consumers' perception of the food safety factors. Meanwhile, 27.6% of the respondents recognize food additives in rice and 22.7% in vegetables. These results indicate that in contrast to the Japanese case, food additives are recognized as a second important factor decreasing food safety.

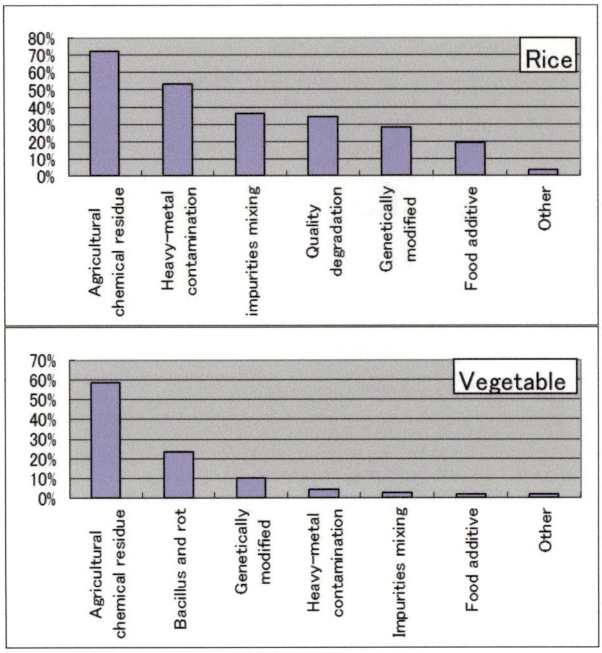

Figure 10.5 Risk factors perceived as decreasing food safety in Japan

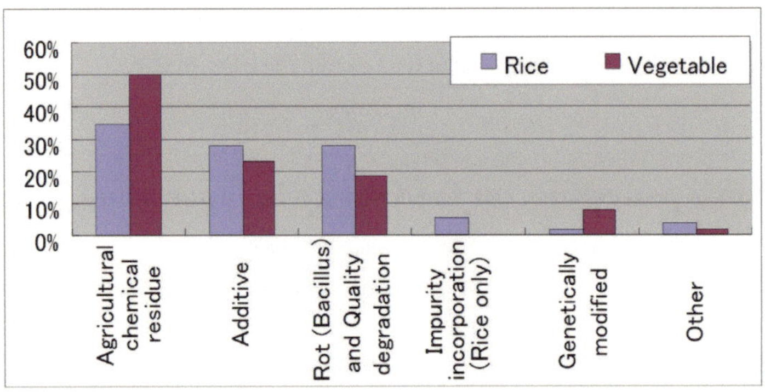

Figure 10.6 Risk factors perceived as decreasing food safety in China

The survey in Japan also showed that 75.7% of consumers recognize cultivation at the farm level (in terms of operation and place) as a key risky process that decreases the safety of imported vegetables. Some 61.3% recognize this issue in domestic vegetables and 53.2% in rice, which is mainly a domestic commodity. Some 66.7% recognize food processing as a key risky process in rice, 47.7% in domestic vegetables, and 62.2% in imported vegetables (Figure 10.7).

For imported vegetables, transportation is also recognized as an important risky process. The results of the surveys in China and in Japan indicated that risk management in production and processing is important for improving consumer trust in the safety of food (Figure 10. 8).

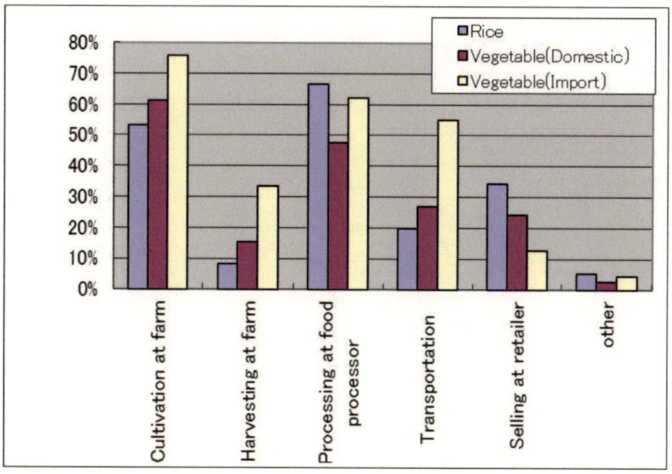

Figure 10.7 Risky processes (operation and place) perceived as decreasing food safety in Japan

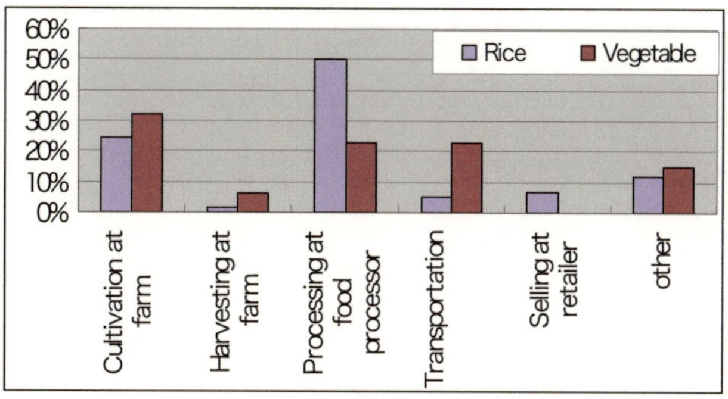

Figure 10.8 Risky processes (operation and place) perceived as decreasing food safety in China

Furthermore, the survey in Japan also showed that more than 95% of the respondents feel that domestic rice (99.1%) and vegetables (95.5%) are safe (Figure 10.9). On the other hand, more than 80% of the respondents feel that imported rice (82.0%) and vegetables (71.2%) are unsafe. The survey in Japan by the Cabinet Office (2008) surveyed the consumers on whether they selected domestic products or imports when buying food. Some 89.0% of the respondents chose domestic food and only 0.5% chose imported food. Some 89.1% of those who selected domestic farm products identified food safety as a selection criterion, along with quality (56.7%), freshness (51.6%), and taste (28.0%). On the other hand, 80.0% of those who selected imported farm products answered that the most important food selection criterion is price, followed by freshness (26.7%) and safety (20.0%). The rate of food self-sufficiency (supply calorie based) of Japan is 40%. Further, Japan imports a huge amount of food from foreign countries every year. Therefore, it is necessary to improve the safety of not only domestic food, but also imported food.

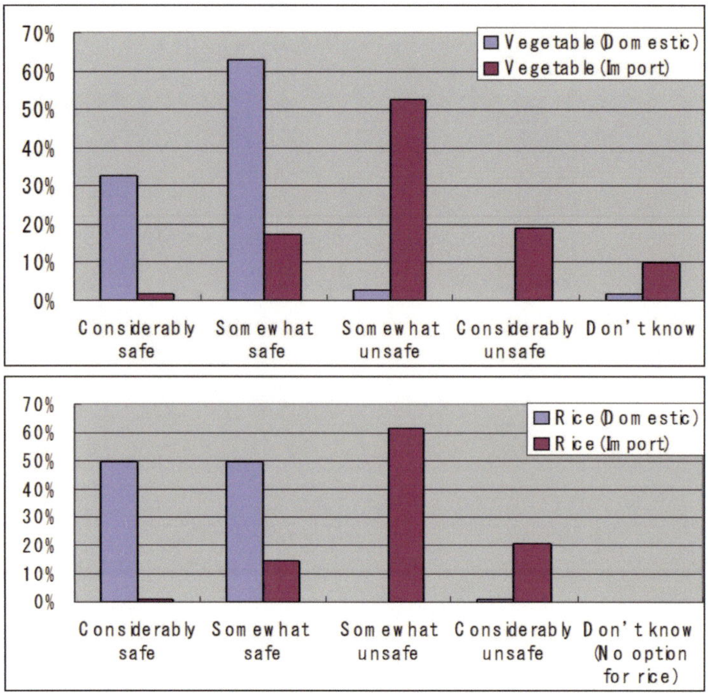

Figure 10.9 Consumers' perceptions on the safety of rice and vegetables in Japan

10.6 Risk management at the farm level in Japan and China

10.6.1 GAP as a risk management approach at the farm level

Good agricultural practices (GAP) and good manufacturing practices (GMP) are sets of principles, regulations, and technical recommendations applicable to production, processing, and food transport, and that address human health care, environmental protection, and improvement of conditions of workers and their families [6]. GAP are a risk management approach at the farm level for safe food production through sustainable farming with low negative environmental impacts. There are various types of GAP around the world. GlobalGAP, formerly known as EUREPGAP, is a nongovernmental GAP.

EUREPGAP was established in 1997 as an initiative by retailers belonging to the Euro-Retailer Produce Working Group (EUREP). Over the last 10 years, as global trading increased, a growing number of producers and retailers around the world began to adopt the practices (Figure 10.10). GlobalGAP are the most acceptable GAP in the world; as of April 2008, 81000 producers from 80 countries and regions are GlobalGAP-certified. Among them, 271 are from the People's Republic of China and six from Japan. In 2011, GlobalGAP-certified producers around the world reached 112600.

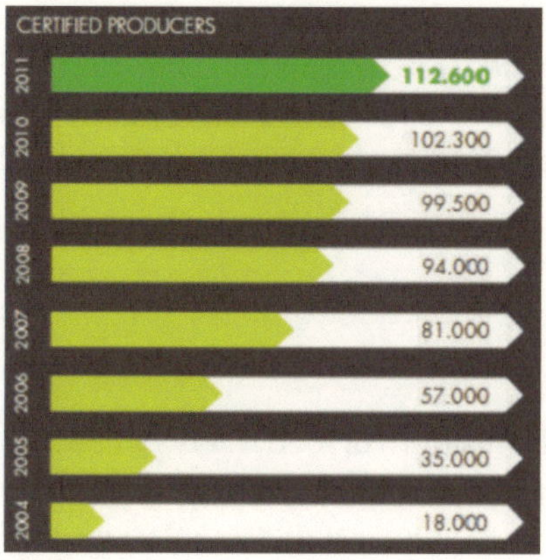

Source: GlobalGAP (2012) [7]

Figure 10.10 GlobalGAP-certified producers

10.6.2 GAPs in Japan and China

The Japan Good Agricultural Initiative (JGAI), established by a group of producers, began its activities, including the extension of GAP, in April 2005. The society, JGAP, was established in November 2006 as an incorporated nonprofit organization, and developed the JGAP Code for vegetables, fruits, and grain. The JGAP has now completed a complete benchmarking procedure, which is fully recognized as a GlobalGAP Integrated Farm Assurance (IFA) Version 3.0 equivalent. As of October 2008, 87 producers, including one in Korea, had obtained this certification. In July 2012, JGAP-certified producers reached 235, including two producers in Korea and Thailand.

In China, the Certification and Accreditation Administration of the People's Republic of China (CNCA) established the China Good Agricultural Practices (ChinaGAP) benchmark system based on the EUREPGAP. On December 31, 2005, the CNCA published the ChinaGAP national standards GB/T20014.1-20014.11-2005 and ChinaGAP rules, both of which apply to grain, fruits, vegetables, dairy cows, beef cattle, sheep, pigs, and poultry. Some 347 enterprises have applied for the ChinaGAP certification, and as of June 2008, 230 enterprises have acquired it [8]. In 2012, the ChinaGAP had certified a total of 108023 producers [9]. This tendency may be because the ChinaGAP is strongly promoted by the government for the export of agricultural products, while the JGAP is mainly promoted by a private organization because Japan's exports of farm products are not as large.

10.6.3 Appropriate agro-chemical use at the farm level

Appropriate agro-chemical use is a major element of GAP. For example, the JGAP (Version 2.1 for vegetable and fruits) has 129 control points, 46 (36%) of which are related to appropriate agro-chemical use. Furthermore, consumers feel that agro-chemicals are a major risk factor decreasing the safety of food in both Japan and China. An information and communication technology-based (ICT-based) decision support system for the risk management of agro-chemical use has been developed, which has been used by more than 47000 farmers as of October 2012 in Japan [10]. This risk management system can be integrated into the food traceability system, and this integration contributes to increased food safety. The system is expected to be introduced in China in order to decrease the risk of inappropriate use of agro-chemicals.

10.7 Concluding remarks

This chapter first reviewed the histories of food safety policy in Japan and China briefly. Next, it clarified the current statuses of the food traceability systems in both countries, and then analyzed consumer perception on food safety from various aspects. Finally, it described the current statuses of risk management at the farm level (i.e., GAP) in both countries.

The authors' various surveys in Japan and China indicated that the recognition rate of the food traceability system by consumers is similar in both countries. The surveys indicated that many consumers do not trust the information provided by food traceability in both countries significantly. This lack of trust is therefore a key issue relating to food traceability. The authors' surveys in both countries showed that both Chinese and Japanese consumers recognize agro-chemical residues as a key risk factor decreasing food safety. Further, the surveys showed that most consumers in both countries recognize cultivation at the farm level (operation and place) as a key risky process decreasing food safety. This result implies that risk management at the farm level, including through GAP, is key to promoting consumer trust in food safety. An integration of the food traceability system and the risk management system at the farm level is therefore essential for food safety. To establish the academic basis of the development of such a system, further researches on risk management at both farm level and regional level, as well as consumer perceptions of food safety, are needed.

References

[1] Min S., Liu L., Wang Z., Nanseki T. Consumers' attitudes to food traceability system in China: evidences from the pork market in Beijing, Journal of the Faculty of Agriculture, Kyushu University 2008; 53(2): 569-574.

[2] Xu Y., Nanseki T., Zeng Y. Food safety issues and consumer perception in China. Nanseki T. [Ed] Food risk and safety in East Asia, Agriculture and Forestry Statistics Publishing Inc. 2010: 101-119 (in Japanese).

[3] goo Research. General results on residents' opinionsregarding food traceability, 2007: http://research.goo.ne.jp/database/data/000665/ (in Japanese).

[4] The Ministry of Agriculture, Forestry and Fisheries of Japan. Food industrial trend investigation 2008: http://www.maff.go.jp/toukei/sokuhou/data/syokuhin2007/syokuhin2007.pdf (in Japanese).

[5] The Cabinet Office. Public opinion poll concerning the role of food, agriculture, and farm village 2008: http://www8.cao.go.jp/survey/h20/h20-shokuryou/index.html (in Japanese).

[6] Food and Agriculture Organization.Guidelines:Good Agricultural Practices for Family Agriculture 2007: http://www.fao.org/docrep/010/a1193e/a1193e00.htm.

[7] GlobalGAP 2012: http://www.globalgap.org/.

[8] Cui, Y. The formation and development of good agricultural practices (GAP), Proceedings of Food Safety and Risk in Agriculture, Resources, and Environment 2008: 47-51 (in Chinese).

[9] Certification and Accreditation Administration of China. Information system of the certified food and agro-products in China 2012: http://ffip.cnca.cn/ffip/publicquery/certSearch.jsp (accessed on Oct. 16, 2012).

[10] Nanseki, T., Kimura H., Hiraishi T., TakahashiT. Development of a risk management system for agricultural chemical use,Agricultural Information Research 2006; 14(3): 207-226 (in Japanese).

[11] Nanseki, T.,Yokoyama K., JAPAN: Improving food safety amongst food operators, Ian G.Smith and Anthony Furness Ed. "Food Traceability Around the World"(in conjunction with the Global Food Traceability Forum) 2008; Vol. 1, p. 46-65.

Index